MEMBRANES

STRUCTURE AND FUNCTION

Sixth FEBS Meeting

Volume 19
METABOLIC REGULATION AND ENZYME ACTION

Volume 20
MEMBRANES Structure and Function

Volume 21
MACROMOLECULES Biosynthesis and Function

FEDERATION OF EUROPEAN BIOCHEMICAL SOCIETIES
SIXTH MEETING, MADRID, APRIL 1969

MEMBRANES
STRUCTURE AND FUNCTION

Volume 20

Edited by

J. R. VILLANUEVA

Department of Microbiology
University of Salamanca
Salamanca, Spain

F. PONZ

Department of Physiology
University of Navarra
Pamplona, Spain

1970

ACADEMIC PRESS · London and New York

ACADEMIC PRESS INC. (LONDON) LTD.
Berkeley Square House
Berkeley Square
London, W1X 6BA

U.S. Edition published by
ACADEMIC PRESS INC.
111 Fifth Avenue
New York, New York 10003

Copyright © 1970 by the Federation of European Biochemical Societies

All Rights Reserved

No part of this book may be reproduced in any form
by photostat, microfilm, or any other means, without
written permission from the publishers

SBN: 12-722150-6
Library of Congress Catalog Card Number: 76-117134

Printed in Great Britain by
Spottiswoode, Ballantyne and Co. Ltd
London and Colchester

Preface

This volume contains the papers presented at the Symposium on "Membranes: Structure and Function", held on 9th April 1969 during the Sixth Meeting of the Federation of European Biochemical Societies at Madrid, Spain. The paper by Dr. W. A. Hamilton was added to this publication at the request of the symposium organizers.

The symposium was organized by W. Wilbrant and J. R. Villanueva. We wish to thank Professor Wilbrant, for his efforts and advice as organizer, and Professors R. A. Peters and M. R. J. Salton, who took the chair during Session a on "Structure of Membranes", as well as Professors H. N. Christensen and F. Ponz who acted as chairmen in Session b on "Transport Mechanisms".

This symposium presents a cross-section of the research now in progress in this field of membranes. Studies on the structure and function of cell membranes over the last few years have led to an increased understanding of the many aspects of these components. In addition to the reviews on the structure and functions of cell membranes in plant and animal cells this volume also presents new experimental evidence contained in papers dealing with the actual composition of cell membranes, their ultrastructure, model organization and chemistry. Other aspects of the subjects are also included such as the effect of some active compounds on bacterial membranes. It is hoped that this symposium will stimulate further work in this very active field of research and give new ideas to future experimentation.

J. R. VILLANUEVA
F. PONZ

April 1970

Contents

PREFACE v

Session a: Structure of Membranes

INTRODUCTORY REMARKS. By R. A. Peters 1
CELL MEMBRANES: THEIR STRUCTURE AND FUNCTIONS. By M. R. J. Salton 3
INTEGRATION OF STRUCTURAL AND BIOCHEMICAL APPROACHES IN THE STUDY OF CELL MEMBRANES. By J. B. Finean and R. Coleman 9
DISASSEMBLY OF THE INNER MITOCHONDRIAL MEMBRANE INTO DIFFERENT SUBMITOCHONDRIAL PARTICLES. By E. Santiago, J. J. Vázquez, J. Eugui, J. M. Macarulla and F. Guerra 17
MOLECULAR ANATOMY OF A MEMBRANE. By J. S. O'Brien . . . 33
PROPERTIES OF THE PURIFIED CYTOPLASMIC MEMBRANE OF YEAST. By Ph. Matile 39
THE RESOLUTION AND PROPERTIES OF SOME MAJOR COMPONENTS OF *Micrococcus lysodeikticus* CELL MEMBRANES. By E. Muñoz, M. R. J. Salton and D. J. Ellar 51
BIOCHEMISTRY OF THE BACTERIAL WALL PEPTIDOGLYCAN IN RELATION TO THE MEMBRANE. By J. M. Ghuysen and M. Leyh-Bouille 59
THE MODE OF ACTION OF MEMBRANE-ACTIVE ANTIBACTERIALS. By W. A. Hamilton 71

Session b: Transport Mechanisms

INTRODUCTION. By H. N. Christensen 81
PHOSPHORYLATIVE TRANSPORT OF SUGARS IN *E. coli*. By G. Gachelin and A. Kepes 87
PRESENT SITUATION IN THE IDENTIFICATION OF MEMBRANE TRANSPORT PROTEINS IN SINGLE CELLS. By A. Kotyk . . 99
REACTIONS AND INTERACTIONS IN INTESTINAL SUGAR TRANSPORT. By R. K. Crane 109

SUCRASE AND SUGAR TRANSPORT IN THE INTESTINE: A CARRIER-LIKE SUGAR BINDING SITE IN THE ISOLATED SUCRASE-ISOMALTASE COMPLEX. By G. Semenza 117

EFFECT OF PHLORETIN AND PHLORIZIN ON SUGAR AND AMINO ACID TRANSPORT SYSTEMS IN SMALL INTESTINE. By F. Alvarado 131

AUTHOR INDEX 141

SUBJECT INDEX 147

Introductory Remarks

R. A. PETERS

Department of Biochemistry, University of Cambridge, Cambridge, England

I consider it a great honour to be invited to open this Session because the subject is so important. We are privileged to have this opportunity of listening to a distinguished group of experts in this field, in which there is much current interest. Let us not think, however, that this interest is only a recent one. It is true that during the era of the "bag of enzymes" theories of the living cell, the question of permeability was relegated to a very secondary place; in fact, it was almost regarded as a last interpretation available to weak biochemistry! However, anyone who has the time to look at the chapter in the book "General Physiology" by Bayliss (1917) will realize that the problems of semi-permeable membranes in cells were recognized as a challenge by powerful minds in the field, much attention being given to the subject in the period 1890 to 1912. Out of this came some interesting models of semi-permeable membranes, as well as the Overton-Meyer theory of anaesthesia with its emphasis upon the necessity of lipoid solubility for penetration of tissues by anaesthetics. The other hypotheses available then now seem crude. For instance, it was held that calcium salts reduced abnormal leakage from tissues. In this connection in 1916/1917, faced as we were by the dire effects upon lungs of gassing with chloropicrin, we solemnly tried the injection of calcium salts in the hope of reducing the extravasation of fluid in the lung of goats; of course it did not work. I mention it only to show how much our thoughts were dominated by the idea that the alveolar membrane in the lung was a passive structure, merely letting gases in and out by simple diffusion. J. S. Haldane always believed that, *in extremis*, the lung could secrete oxygen, for example in Pikes Peak, but it was not easy to put his views beyond argument. Noone then had an idea that the presence of mitochondria, etc. controlled cell activities in such membranes, but we had been made very nervous by the rightful criticism of W. B. Hardy (1899) of the effect of fixatives in inducing artefactual structures in apparently homogeneous, gel-like structures. We were not really rescued from this situation until the advent of phase contrast microscopy and of the electron microscope; and now we have good new techniques, and also a group of histologists thoroughly alerted to the nature of histological artefacts.

From the late 1920's, thought upon the structure of membranes has been somewhat dominated by the bimolecular lipid theory of the cell membrane, arising from Gorter and Grendel and advanced by Danielli and Davson. This theory has been useful; it is consistent with the Overton-Meyer theory and with the observed low interfacial tension between cells and their environment. I have never been personally happy with this model nor with the "bag of enzymes" theories; though the latter gave us our present wide knowledge of the chemistry of biochemical entities, it always seemed to me to be philosophically weak. I think that it meant that the implications of the work on orientation in surfaces and on monomolecular films was not really understood, or the extreme microhetereogeneity of living cells. As long ago as 1929, I had pointed out the need for some kind of fluid cytoskeleton, a mosaical structure involved in the direction of cell activities and their coordination, the consequences of which must mean that cell enzymes were under the control of the whole cell; furthermore, I felt that the control must require receptor groups of a protein nature on the outside of the cell. The presence of internal structures, organelles and endoplasmic reticulum has now been made so clear that it is unnecessary for me to say more of them. We shall hear much about these in the following articles. I am glad to see that some recent research is being directed to the problem of structural protein running as ? lipoprotein through the cell membrane. We ought soon to have solid information, because we have many active minds in the field, and, as you will hear, a vastly superior array of techniques which was not before available to interested workers.

Cell Membranes: Their Structure and Functions

M. R. J. SALTON

*Department of Microbiology, New York University
School of Medicine, New York, U.S.A.*

The existence of well-defined boundaries to cells and organelles within cells has been recognized by microscopic anatomists for a very long time. The functions of such cell-surface boundaries or membranes in transport and permeation processes has been well established and the possible arrangement of the two major classes of chemical substances, the proteins and lipids, was suggested as long ago as 1925 by Gorter and Grendel (1925) and by Danielli and Davson (1935). The resolution of the structure of cell membranes by electron microscopy of thin sections of a great variety of cells, intracellular organelles and membrane systems suggested the existence of a rather uniform type of structure, the "unit membrane" (Robertson, 1959). Moreover, the double-track feature of the membrane profiles seen in thin sections of cells of animal, plant and microbial origins coincided remarkably well with the bimolecular leaflet model proposed by Danielli and Davson (1935). Sjöstrand (1953) was one of the first investigators to suggest that the Danielli-Davson model for the plasma membrane may also be applicable to cytomembranes and a variety of other cellular membranes. However, the double-track or trilamellar appearance of the membrane profile in fixed, sectioned material gives an over-simplified impression of the membrane structure, and the simple lipid-protein "sandwich" of the unit membrane is being replaced by "granulo-fibrillar", "globular" or "tripartite" structures possessing more complex molecular arrangements.

The problems of interpreting the molecular architecture of cell membranes on the basis of the electron opacity or variations in the electron density of the structures seen in cells and tissues fixed with osmium tetroxide or potassium permanganate and stained with heavy metal salts are tremendous, and Sjöstrand (1967) has discussed various aspects of this. As Korn (1968) has pointed out, in order to interpret electron microscopic images in molecular terms it would be necessary to establish what is responsible for the adsorbed and scattered electrons and to know which molecules in the membrane bind the atoms of Os, Mn, U and Pb. Indeed, there is very little information available on the chemistry of the interaction of $KMnO_4$, OsO_4, and glutaraldehyde with specific groupings of the membrane proteins and lipids. Some progress has been made in studying the *in vitro* reactions of osmium tetroxide with lipids (Korn, 1968). Glutaraldehyde, which is active in cross-linking of proteins and often regarded as an excellent fixative, cannot yield information about the location of lipids in the membranes as the unchanged lipids are completely extracted during dehydration in ethanol (Korn and Weisman, 1966). The apparent loss of outer mitochondrial membranes under some conditions of fixation with glutaraldehyde, the absence of any change in the profile image of mitochondrial membranes subjected to prior extraction of lipids with acetone (Fleischer *et al.*, 1967), and the loss and destruction of lipids on exposure to OsO_4 all emphasize the difficulties of assigning any specific molecular orientation of interactions to the profiles of membranes seen in sectioned material examined in the electron microscope.

The freeze-etching technique which avoids some of the problems of fixation, dehydration in organic solvents, and chemical degradation of membrane components is not completely divorced from ambiguity of interpretation (Moor and Muhlethaler, 1963; Branton, 1966). The membranes examined by Branton (1966) had "the best of both worlds" in that the fractured faces were organized in part as a bilayer and in part as globular subunits. An extension of this powerful method of preparing membranes for topographical studies in the electron microscope to the area of identification of membrane-specific enzymes or functions (e.g. active transport proteins) and/or characteristic groupings within the protein and phospholipid molecules, should give us a more precise understanding of the molecular architecture of the membrane in a state of preservation much closer to that of its native form.

Although the pictorial representation of cell membrane structure elucidated by electron microscopy has been plagued with interpretational problems, I would hesitate to imply that the results emerging over the past decade have not been worthwhile. It is clear now that we understand the problems of interpretation much better and the consequences of "immobilizing" and staining a dynamic structure for examination in the electron microscope. It is likely that the electron microscope will play a major role in establishing the fine structure of cell membranes by enabling us to locate specific functional proteins on the membranes and investigating their distribution during biogenesis of these important structures. The localization of antigenic proteins, be they functional or structural components of the membrane, by the use of ferritin-labelled antibodies or the peroxidase-antibody conjugated complex, should yield valuable information about the structural arrangement of the membrane. These methods, together with specific enzyme staining techniques, should throw light on the problem of the mechanism whereby a membrane can conserve its "polarity" or asymmetry while performing such functions as active transport, secretion of proteins, etc. Apart from the localization of blood group substances on the outer surfaces of erythrocyte membranes and ATPase on their inner surfaces there is little precise information about the location and distribution of other "structural" or functional entities of the cell membrane.

The development of methods other than electron microscopy has added much to our knowledge of membrane structure and the kinds of molecular interactions which hold them together. Early X-ray diffraction studies yielded data which showed a good correspondence to the density spacings seen in the electron microscopic profiles. X-ray diffraction data have been used to assess the minimum thickness of each protein layer in myelin. Recent low-angle X-ray diffraction studies of erythrocyte ghosts have suggested an overall thickness of about 110 Å but attempts to resolve any regularity of distribution of discrete macromolecular units on examination of hydrated membranes have not been successful (Finean et al., 1966).

Considerable progress has been made over the past few years in studying the conformation of membrane proteins and establishing the nature of their interactions with lipids. Thus the configuration of peptide chains in membrane proteins has been investigated by the methods of infrared and NMR spectroscopy and by optical rotatory dispersion and circular dichroism (for original references see Chapman, 1968). The results of these recent independent investigations have indicated: (1) that membranes of animal and bacterial origins possess about 20-50% of their proteins in the α-helix configuration; (2) that there are strong interactions between α-helical regions of adjacent proteins; and (3) evidence for non-ionic, hydrophobic bonding between lipids and proteins. The evidence that membranes owe their structural integrity to weak hydrophobic interactions between lipids and proteins as well as protein-protein interactions has led to a complete revision of the biomolecular leaflet as an acceptable model for all membranes. To accommodate this new evidence, the latest membrane models place the hydrophilic groups of the phospholipids and proteins at the aqueous interfaces, the whole structure being "stabilized" by non-polar interactions between proteins and lipids as well as hydrophobic interactions within the two classes of membrane components, i.e. protein-protein, lipid-lipid. At the present time there is no quantitative information on the relative amounts of these different types of molecular associations in membranes.

Chemical analysis of isolated cell membranes has established that all biomembranes are composed of lipid and protein, although there are wide differences in the ratios of protein to lipid, varying from 0.25 for myelin, to 3.0 for bacterial membranes (Korn, 1968). Phospholipids account for a major portion of the total lipid content of cell membranes, while cholesterol ranges all the way from being the other principal constituent of the lipid of myelin and erythrocyte membranes, to being completely absent in all bacterial membranes with the exception of members of the Mycoplasma group. More and more information about the nature of the membrane proteins is now becoming available and Mazia and Ruby (1968) have suggested that they be classified as "tektins" as they constitute a distinctive group of proteins with respect to amino acid composition. Examination of dissociated membranes and solvent extracted membranes for their protein components has usually revealed a complex mixture separable by electrophoresis in polyacrylamide gels. The interpretation of the resolution of membrane proteins in disc gels containing dissociating agents such as sodium dodecyl sulfate or urea is not without its difficulties since incomplete dissociation could lead to multiple banding. Lipoproteins have been released from some membrane preparations and this, together with studies of the effectiveness of various solvents in removing lipids, suggest that the firmness of binding between individual proteins and lipids may vary. No definitive evidence for covalent bonding between membrane lipid and protein is available at present. Virtually all of the phospholipid in the membrane of *Micrococcus lysodeikticus* can be removed from the protein by electrophoresis with deoxycholate (Salton and Schmitt, 1967), a procedure which would disrupt weak associations in the membrane without breaking covalent bonds.

Despite the apparent similarity of all membrane structures implied by the appearance of the electron microscopic profiles, it has become abundantly clear that both their structural and functional properties are far from being uniform. Thus, membranes may range from the least biochemically active, relatively inert myelin sheath with its "insulator" functions, to the complex respiratory functions packaged into the inner and outer mitochondrial membranes. Membranes such as those of erythrocytes possess selective transport mechanisms as well as a variety of enzymes which apparently include glycolytic enzymes. Liver plasma membranes are a reservoir of some twenty or more enzymes. Because of the absence in bacteria of specialized organelles performing specific functions associated with nuclear division, respiratory activity, protein synthesis on membranous endoplasmic reticulum, the relatively undifferentiated membrane systems of the bacterial cell are required to supply the needs for these cellular activities. Thus, bacterial membranes possess functions of the plasma membrane (active transport of sugars, amino acids, etc.) as well as mitochondrial functions, those needed for nuclear development and separation during division, in addition to being the site of enzymatic assembly of the complex external peptidoglycan cell-wall structure.

The isolation and characterization of enzymes and specific transport proteins from membranes and from the periplasmic regions of bacteria offers an exciting prospect in studying the mechanisms of certain membrane functions. There is little doubt that the next era of membrane research will see some of these functions described in rather specific terms of protein conformations and in so doing we shall have learned a great deal about membrane structure and membrane function. Finally, the mechanism of biogenesis and *in vivo* assembly of cellular membranes is little understood and its elucidation will reveal much about the molecular architecture and regional specificity of membranes.

It is appropriate that the papers selected for this Symposium span a variety of membrane structures and functions and we are all looking forward to hearing the exciting details that these papers will contribute to one of the major areas of biology.

REFERENCES

Branton, D. (1966). *Proc. natn. Acad. Sci. U.S.A.* **55**, 1048.
Chapman, D. (ed.) (1968). "Biological Membranes", Academic Press, London and New York.
Danielli, J. F. and Davson, H. (1935). *J. cell. comp. Physiol.* **5**, 495.
Finean, J. B., Coleman, R., Green, W. A. and Limbrick, A. R. (1966). *J. Cell Sci.* **1**, 287.
Fleischer, S., Fleischer, B. and Stoeckenius, W. (1967). *J. Cell Biol.* **32**, 193.
Gorter, E. and Grendel, F. (1925). *J. exp. Med.* **41**, 439.
Korn, E. D. (1966). *Biochim. biophys. Acta.* **116**, 317.
Korn, E. D. (1968). *In* "Biological Interfaces: Flows and Exchanges", p. 257, Little, Brown and Company, Boston.
Korn, E. D. and Weisman, R. A. (1966). *Biochim. biophys. Acta* **116**, 309.
Mazia, D. and Ruby, A. *Proc. natn. Acad. Sci. U.S.A.* (1968). **61**, 1005.
Moor, H. and Muhlethaler, K. (1963). *J. Cell Biol.* **17**, 609.
Robertson, J. D. (1959). *Biochem. Soc. Symp.* **16**, 3.
Salton, M. R. J. and Schmitt, M. D. (1967). *Biochem. Biophys. Res. Commun* **27**, 529.
Sjöstrand, F. S. (1953). *Nature, Lond.* **171**, 31.
Sjöstrand, F. S. (1967). *Protides biol. Fluids* **15**, 15.

Integration of Structural and Biochemical Approaches in the Study of Cell Membranes

J. B. FINEAN and R. COLEMAN

Department of Biochemistry, University of Birmingham, Birmingham, England

A variety of well-tried methods of structural analysis have now been applied to intact membranes and a number of structural parameters established which do not, however, provide a precise or unique molecular picture of a membrane (Finean, 1969—recent review of structural data).

Electron micrographs of tissue sections have emphasized a lamellar aspect and provided approximate dimensions for a very wide variety of membranes. Electron micrographs of other types of membrane preparations, such as those prepared by negative staining or freeze etching, have suggested that the lamellar framework may have a granular substructure and some experiments have indicated a complete subdivision of some membranes into repetitive lipoprotein subunits.

Low angle X-ray diffraction patterns of hydrated membrane systems have been interpreted in terms of an electron density profile through the membrane layer which is indicative of a lamellar arrangement of the principal molecular constituents. In two cases (chloroplasts and retinal receptors) diffraction data relating to a granular substructure within the plane of the membrane have also been recorded but in both cases they relate to a superficial granular layer rather than to a complete subdivision of the membrane.

X-ray diffraction data have indicated that the lipid components of the membrane form a liquid crystalline phase but nuclear magnetic resonance (NMR) spectra suggest that there is some restriction of the movement of hydrocarbon chains despite the overall high degree of disorder in the hydrocarbon phase. Measurements of optical rotatory dispersion (ORD) and of circular dichroism (CD) indicate that a high proportion of the protein is in an α-helical configuration and the remainder predominantly random coil, but there are perturbations of the spectra which are attributed to the molecular environment and may reflect in part a hydrophobic interaction of protein with the hydrocarbon regions of the lipid.

With all these approaches there appears to be some ambiguity or impreciseness in the interpretations of the data from this type of complex molecular system, and several groups of workers have sought to simplify or at least to carry out some significant modifications of the system in the hope of reducing the ambiguity or adding to the detailed significance of the structural data.

A great variety of agents, such as specific degradative enzymes, solvents, detergents, chelating agents and fixatives, have been used to modify a variety of membranes, and a wide range of observations have been reported, but studies have rarely been systematic or comprehensive and a full review would be complex and perhaps confusing. Instead, we have chosen to discuss just two modifying agents, trypsin, to modify membrane protein, and phospholipase C, to modify membrane lipid, and to emphasize studies of two membranes, erythrocyte ghosts, a plasma membrane, and muscle microsomes which are derived predominantly from the endoplasmic reticulum of muscle cells.

MODIFICATION BY TRYPSIN

Trypsin cleaves peptide bonds adjacent to basic amino acid residues such as arginine and lysine. The maximum amount of peptide bond material liberated from a variety of membrane preparations by trypsin has varied considerably (in our own experiments from 15% to 50% of the total protein) but in no case has the protein removed been identified with a specific membrane protein and no attempt has been made to study (e.g. by ORD or CD measurements) changes in protein configuration associated with tryptic action on membranes.

Studies of the action of trypsin on cell membranes so far have included observations on the effects on permeabilities, enzyme activities, the appearance in electron micrographs, and the X-ray diffraction patterns. Fairly extensive studies have been made of the effects on haemoglobin-free erythrocyte ghosts, liver plasma membranes and muscle microsomes.

Trypsin releases up to 50% of the total peptide from haemoglobin-free erythrocyte ghosts (S. Knutton, unpublished data). The main effect on the morphology of the membranes, as seen in electron micrographs of sectioned preparations, is a break-up of the collapsed membranous sacs which characterize the normal membrane preparation to give large independent segments of membrane. The membranes still have a trilamellar structure but when the preparation is condensed for X-ray diffraction studies it becomes evident that the overall thickness of the membrane has decreased by about 40% and that the decrease represents removal of material predominantly from the outer surface of the membrane. Electron micrographs of negatively stained preparations of trypsin-treated red cell membranes, published by Marchesi and Palade (1967), also indicated fragmentation of the ghosts and revealed a short filamentous component superimposed on the membrane fragments. Such preparations were found to have lost 80% of their Na^+-K^+-Mg^{2+}-stimulated ATPase activity, but only 15% was lost when ATP was present during the tryptic digestion. Studies of the effects of trypsin on intact red cells showed the Na^+-K-Mg^{2+} ATPase to be unaffected by this modification, but cholinesterase activity was lost (Herz et al., 1963). This would seem to indicate a localization of the ATPase activity on the inside of the membrane so as not to be accessible to trypsin when applied to the intact cell. On the same reasoning, cholinesterase would appear probably to be located on the outside.

When liver plasma membranes were treated with trypsin, Benedetti and Emmelot (1968) found the overall thickness of the trilamellar unit, characteristic of the membrane in sectioned preparations, to be reduced from a normal value of about 80 Å to about 60 Å. Electron micrographs of negatively stained membranes showed marked changes in surface structure as compared with control preparations.

Tryptic action on red cell membranes has also been studied by X-ray diffraction methods and the most striking change has been in the low angle diffraction pattern of the dried preparation. Dehydration of these membranes normally results in the separation of a part of the membrane lipid to form independently diffracting lipid phases but a residual lipoprotein structure remains. The lipids give reflections at 53 Å and 43 Å and a band at 80 to 90 Å is identifiable with the residual lipoprotein. Following trypsin treatment and drying, the preparation no longer gave the lipoprotein band, and to the lipid bands at 53 Å and 43 Å, which probably represent mainly phospholipid, was added a further reflection at about 34 Å which would be expected to arise from crystallized cholesterol. In effect, the trypsin treatment has so modified the lipid-protein interaction in the membrane that during drying the whole of the lipid component crystallizes out and the cholesterol (possibly by virtue of there being more of it now separated) more readily forms an independent phase.

Several groups of workers have studied the action of trypsin on preparations of muscle microsomes from a variety of viewpoints (Martonosi, 1968b; Inesi and Asai, 1968; Coleman et al., 1969). In all cases the maximum amount of protein removed was 40% to 50% but this was characterized only in terms of lost activities of the membrane vesicles. The loss of ability of the vesicles to accumulate calcium was accompanied by the removal of only a few per cent of the total membrane protein, but complete suppression of Mg^{2+} ATPase activity required the removal of at least 40% of the protein. Electron micrographs of sectioned preparations showed that removal of only small amounts of protein led to a breaking open and to a collapsing of membrane vesicles, whilst maximum protein removal left mainly strands (or plate-like fragments) of material which did, however, retain a trilamellar-type cross-section. Electron micrographs of negatively stained preparations of trypsin-treated membranes (Martonosi, 1968c) again showed a depletion of a superficial particulate structure.

Low angle X-ray diffraction patterns of dried preparations of trypsin-treated muscle microsomes showed only lipid reflections. As in the case of the red cell membrane preparation the residual lipoprotein reflection which normally persists in the dried preparation is not seen in the case of the trypsin-treated material, again indicating that the trypsin treatment has so modified the lipid-protein interaction that all of the lipid crystallizes out when the membranes are dried.

MODIFICATIONS BY PHOSPHOLIPASE C

Phospholipase C (phosphatidylcholine cholinephosphohydrolase (E.C. 3.1.4.3.)) preparations have been observed to cleave a variety of phospholipids (van Deenen, 1964) but in some circumstances there appears to be a preferential attack on diacyl phosphatidylcholines (lecithin) and on sphingomyelin. The cleavage liberates the entire phosphate-containing ionic end group (e.g. phosphoryl choline).

Phospholipase C preparations have been used to modify a variety of membrane preparations and, in the few cases in which quantitative analyses have been reported, from 50% to 80% of the total phospholipid has been hydrolysed. In studies of muscle microsomes Martonosi (1968a) has shown that over 90% of the membrane lecithin (comprising 60% to 70% of the total phospholipid) is hydrolysed by phospholipase C but the other lipids distinguishable on the chromatograms appear to be little affected. In our own recent studies of haemoglobin-free human erythrocyte ghosts, analysis of the lipids by thin layer chromatography has demonstrated that the main target of phospholipase C is lecithin and sphingomyelin with a smaller effect on phosphatidyl ethanolamine and very little effect on the more ionic phospholipids. The lipid hydrolysed by phospholipase C again accounts for about two-thirds of the total membrane phospholipid.

No study has yet been made of the physical state of the residual lipid but optical rotatory dispersion and circular dichroism studies of phospholipase C-treated erythrocyte ghosts (Lenard and Singer, 1968; Gordon et al., 1969) have indicated that the protein conformation is little changed although there is a small shift in the spectrum which still requires final explanation.

Electron micrographs of sections through phospholipase C-treated erythrocyte ghosts (Fig. 1) and muscle microsomes (Finean and Martonosi, 1965) have shown uniformly dense droplets associated with membranous sacs which still retain a trilamellar structure. Such dense droplets have been partially released by mild sonication treatment and separated by density gradient centrifugation. A variety of analyses have indicated that this fraction is rich in diglyceride and cholesterol but it also includes some membrane fragments.

Electron micrographs of negatively stained preparations of phospholipase

STRUCTURAL AND BIOCHEMICAL STUDIES OF MEMBRANES 13

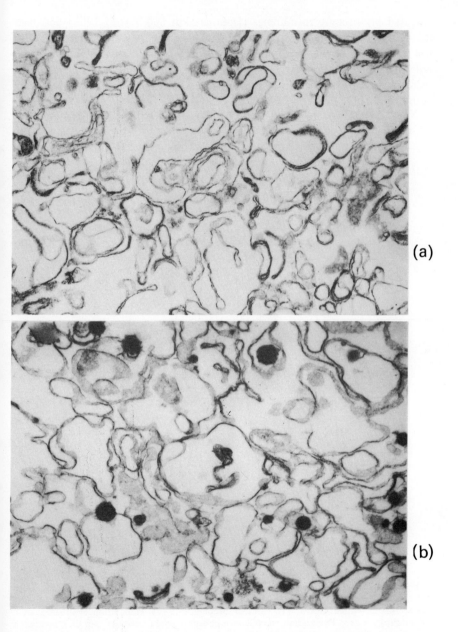

Figure 1. Electron micrographs of haemoglobin-free preparations of human erythrocyte ghosts. OsO_4-fixed. Sections stained with uranyl acetate. (Magnification × 17,000). (a) Control preparation incubated at 37°C for 15 min at pH 7.4; (b) Phospholipase C-treated preparation. Incubation conditions as in legend to Fig. 2.

C-treated erythrocyte membranes (Dourmashkin and Rosse, 1966) and liver plasma membranes (Benedetti and Emmelot, 1968) have revealed large (350 to 500 Å diameter) ring-like structures or defects in the membrane but no molecular interpretation has yet been suggested. Phospholipase C-treated muscle microsomes showed no appreciable modification (Martonosi, 1968c) by this method.

X-ray diffraction studies of phospholipase C-treated membranes have again emphasized marked changes in the diffraction patterns of the dried preparations (Fig. 2). In the case of dried preparations of phospholipase C-treated erythrocyte

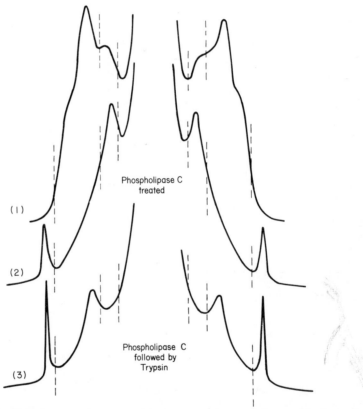

Figure 2. Microdensitometer traces of low angle X-ray diffraction patterns from dehydrated samples of erythrocyte ghosts. Vertical interrupted lines indicate corresponding regions in the three traces. (1) Control preparation; (2) Preparation incubated with phospholipase C; (3) Preparation treated with phospholipase C, washed, and then incubated with trypsin. Phospholipase C incubation, 10:1, membrane protein: enzyme protein (Sigma *Cl. Welchii*, Type I). 15 min at 37°C, pH 7.4, 0.0025 M Ca^{2+}. Trypsin incubation, 50:1, membrane protein: enzyme protein (Sigma, Type III). 20 min at 37°C, pH 7.4. Stopped with trypsin inhibitor (Soya bean, Type IS).

ghosts there are no longer any diffraction bands in the 40 to 60 Å region where one expects to find the phospholipid reflections, but the residual lipoprotein band is enhanced as compared with the normal dried preparation. There are also sharp and intense reflections at 34 Å and 17 Å that are clearly identifiable as arising from crystalline cholesterol. Phospholipase C-treated membranes which were subsequently digested with trypsin and then dried gave a low angle X-ray diffraction pattern which no longer featured a lipoprotein reflection, but there was again a reflection in the 40 to 60 Å region and the cholesterol diffractions were further enhanced. A further treatment with phospholipase C had no appreciable effect on the reflection in the 40 to 60 Å region.

Considerations of chromatographic and diffraction data lead to the conclusion that phospholipase C digests that part of the membrane phospholipid which normally separates from the membrane when it is dried (i.e. the labile lipid) and that this lipid features mainly the less highly charged choline and ethanolamine containing phospholipids. The cholesterol that was associated with this phospholipid in the membrane appears to crystallize out more readily when the phospholipid is hydrolysed to give diglyceride. The phospholipid which remains part of the residual lipoprotein framework when erythrocyte membranes are dried, and which resists the action of phospholipase C, includes the more highly charged molecules such as phosphatidyl serine and the inositol phosphatides. In the case of erythrocyte membranes it is therefore possible to distinguish a more labile lipid component which includes the less polar phospholipids and a more tightly bound lipid component in which the more highly polar lipids are prominent. Such observations could be indicative of an important ionic interaction between lipid and protein in this membrane.

Dried preparations of other mammalian cell membranes also give X-ray diffraction patterns in which residual lipoprotein reflections and reflections representing purely lipid phases can be distinguished. It will be of interest to identify the different categories of lipids in these membranes as well, in order to see if the findings for the erythrocyte membranes have a general significance.

Treatment of isolated membrane preparations with phospholipase C has been shown to inhibit several membrane activities. The Na^+-K^+-Mg^{2+} ATPase of erythrocyte ghosts (Schatzmann, 1962) and of liver plasma membranes (Benedetti and Emmelot, 1968) are suppressed. The Mg^{2+} ATPase of muscle microsomes is completely inhibited by phospholipase C but is fully restored by the addition of some phospholipid preparations (Martonosi, 1968a). These observations appear to have general structural implications but the detailed significance cannot yet be specified.

The general conclusion that must be drawn from the available experimental data on the effects of degradative enzymes on membrane structure and function is that although a large number of observations have been reported they remain largely uncoordinated. There are, however, some promising leads emerging and an extension of combined structural and biochemical studies to a wider variety of membranes should lead to a further clarification of the functional significance of molecular relationships in membranes and to a more detailed description of such molecular relationships.

SUMMARY

Structural and functional data from cell membranes may be extended and their significance enhanced through studies of biochemically significant modifications of the membrane preparations. Trypsin and phospholipase C have been used extensively to modify protein and phospholipid components of membranes. Erythrocyte ghosts and muscle microsomes have provided the most convenient membrane preparations. This paper reviews some of the published data and also emphasizes some preliminary X-ray diffraction and electron microscope studies of membranes degraded by enzymes.

REFERENCES

Benedetti, E. L. and Emmelot, P. (1968). *In* "Ultrastructure in Biological Systems", (A. J. Dalton and F. Haguenau, eds.), Vol. 4, p. 33, Academic Press, New York.
Coleman, R., Finean, J. B. and Thompson, J. E. (1969). *Biochim. biophys. Acta* **173**, 51.
van Deenen, L. L. M. (1964). *In* "Metabolism and Physiological Significance of Lipids" (R. M. C. Dawson and D. N. Rhodes, eds.), p. 155, John Wiley, London.
Dourmashkin, R. R. and Rosse, W. F. (1966). *Am. J. Med.* **41**, 699.
Finean, J. B. (1969). *Q. Rev. Biophys.* **3**, 43.
Finean, J. B. and Martonosi, A. (1965). *Biochim. biophys. Acta* **98**, 547.
Gordon, A. S., Wallach, D. F. H. and Straus, J. H. (1969). *Biochim. biophys. Acta.* In press.
Herz, F., Kaplan, E. and Stevenson, J. H. (1963). *Nature, Lond.* **200**, 901.
Inesi, G. and Asai, H. (1968). *Archs Biochem. Biophys.* **126**, 469.
Lenard, J. and Singer, S. J. (1968). *Science, N. Y.* **159**, 738.
Marchesi, V. T. and Palade, G. E. (1967). *Proc. natn. Acad. Sci. U.S.A.* **58**, 991.
Martonosi, A. (1968a). *J. biol. Chem.* **243**, 61.
Martonosi, A. (1968b). *J. biol. Chem.* **243**, 71.
Martonosi, A. (1968c). *Biochim. biophys. Acta* **150**, 694.
Schatzmann, H. J. (1962). *Nature, Lond.* **196**, 977.

Disassembly of the Inner Mitochondrial Membrane into Different Submitochondrial Particles

E. SANTIAGO, J. J. VÁZQUEZ, J. EUGUI, J. M. MACARULLA
and F. GUERRA*

*Department of Biochemistry, CIB "Felix Huarte", University of
Navarra, Pamplona, Spain*

Morphologically, mitochondria are described as tiny spheres, short rods, or filaments in which the outer membrane, inner membrane, and matrix are present as ultrastructures. Folds of the inner membrane called *cristae mitochondriales* protrude into the interior of the mitochondria as incomplete septum-like structures. On the other hand, plant mitochondria show tubular structures rather than cristae. The matrix, enclosed by the inner membrane, often has a filamentous appearance.

This classical image of the mitochondria, obtained from observations of tissue embedded and sectioned in the conventional way, is notably different from that derived from negatively stained preparations.

Negative staining has brought into view new structures of the mitochondria about which there is no unanimous agreement. Some would interpret these ultrastructures as artifacts produced during the isolation procedure of the membranes (Mitchell, 1967a,b; Sjöstrand et al., 1964).

INNER MEMBRANE ULTRASTRUCTURE

We have presented evidence that inner mitochondrial membranes from rat liver show two different types of structures, tubules and lamellae. The tubules showed the typical projecting subunits, while the lamellae completely lack the projecting subunits.

Isolated mitochondria were subjected to osmotic rupture following the method of Parsons et al. (1965). Inner mitochondrial membranes were obtained using Parsons "low speed pellet" (LSP) as starting material. In order to remove the outer membrane present in this fraction the LSP was washed three times by resuspending it in 0.02 M phosphate buffer, pH 7.4, centrifuging at 1900 x g for 15 min and once more resuspending it in 0.25 M sucrose and centrifuging at

* Permanent address: Centro de Estudos de Bioquimica do Instituto de Alta Cultura, Faculdade de Farmacia, Porto, Portugal.

8500 x g for 10 min. They were then fixed in buffered osmium tetroxide at 3°C, negatively stained (Santiago et al., 1968a), and then examined with a Siemens Elmiskop IA. The objective lens was carefully compensated and equipped with 50 or 30 μ aperture diaphragms; electron micrographs were taken at plate magnifications between 15,000 and 60,000.

Negatively stained mitochondria show an abundance of interior structures. Fig. 1 corresponds to a mitochondrion after hypotonic rupture of its outer

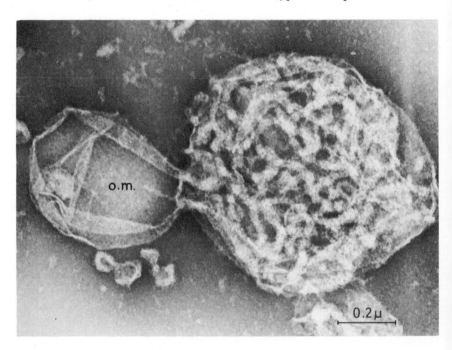

Figure 1. Electron micrograph of a mitochondrion negatively stained, after hypotonic rupture. The outer membrane (o.m.) appears as an empty bag leaving a large quantity of tubular structures to its right.

membrane and before the usual centrifugation to separate inner membranes from the matrix and outer membranes. One can observe the delicate outer membrane (o.m.) in the form of an empty bag leaving behind a large quantity of tubular structures to its right.

In our negatively stained inner membrane preparations we find two types of structures: one tubular and another lamellar. Fig. 2 shows a typical tubular structure with Parsons projecting subunits (Parsons, 1963), first described by Fernandez-Morán (1962) under the name of elementary particles. They cover the entire outer surface of the tubules. Lamellae represent a minor component and they completely lack projecting subunits. Frequently, tubular material is found to join lamellar areas forming long chains of alternating tubular and lamellar links, as is shown in Fig. 3. At times, however, the smallest lamellar links do show projecting subunits.

Occasionally, two or more tubules are seen as continuations of the lamellar structure and at the same time show bifurcation. The point of junction between the tubules and lamellae is often funnel shaped as seen in the lower centre portion of Fig. 4.

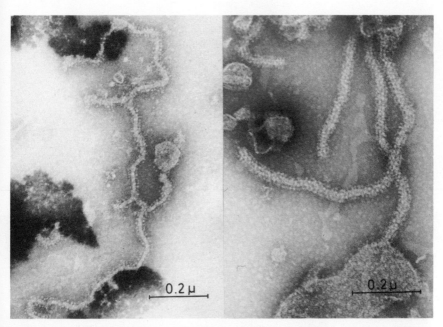

Figure 2. Tubular structure of inner mitochondrial membrane showing projecting subunits. Left: tubule showing several bifurcations. Right: tubules studded with projection subunits over its entire surface area. (Negative staining.)

The tubules are from 200 to 375 Å in diameter and the centre, being less dense, permits the contrast material to penetrate. The length of the tubules varies considerably and at times reaches more than 7 μ.

The tubular component in the inner membrane preparations from rat liver mitochondria is more abundant than the lamellar one.

Negative staining of mitochondrial inner membrane gives a quite different picture from that obtained from ultrafine sections of embedded tissue. The embedded sections show a comparatively low number of cristae whereas negatively stained preparations give a number of lamellar structures and many tubules. Some tubules are found to be longer than the actual length of mitochondria.

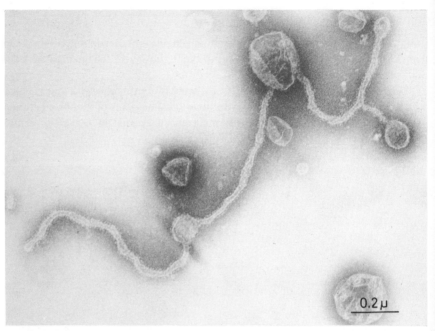

Figure 3. Electron micrograph of inner mitochondrial membrane showing tubules connecting balloon-like structures. (Negative staining.)

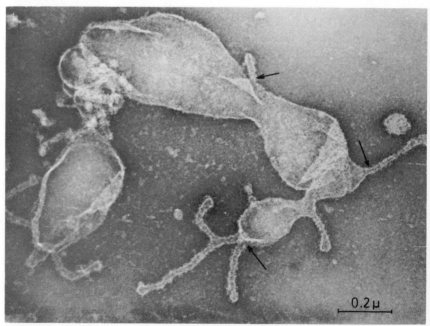

Figure 4. Electron micrograph of inner mitochondrial membrane showing lamellar and tubular structures. The points of junction (arrows) are funnel shaped. (Negative staining.)

The tubular material is definitely such, since in the osmium tetroxide fixation previous to negative staining they are seen lying down on the supporting film and studded with projecting subunits over their entire surface area. This excludes the possible "cristae viewed in profile" interpretation proposed (Fernandez-Morán, 1962; Fernandez-Morán et al., 1964; Stoeckenius, 1963). The fixation with osmium tetroxide does not eliminate the projecting subunits as suggested by Stoeckenius (1963).

One of the most interesting aspects about these tubular structures, and even about projecting subunits themselves, is whether they are structural elements really existing as such in the mitochondria or, on the contrary, are only artifacts as Mitchell (1967a,b) and Sjöstrand et al. (1964) point out. The main reason for not accepting that such structures are present in intact mitochondria is that they have not yet been convincingly demonstrated in sections of embedded tissue. At the same time it is difficult to picture as an artifact such an orderly and regular structure and one that is unique to inner mitochondrial membranes. It is not possible that their formation is induced by phosphotungstate on making the contrast, since the membranes were previously fixed with osmium tetroxide. They would have to be formed during the short time of hypotonic rupture.

EFFECT OF ASCORBATE ON ISOLATED INNER MITOCHONDRIAL MEMBRANES

Studies carried out in our laboratory (Santiago et al., 1968b) have shown that isolated mitochondria incubated in the presence of ascorbate undergo marked changes in their phospholipid composition. Both phosphatidyl choline and phosphatidyl ethanolamine decrease to a large extent during incubation. These phospholipid changes are parallel to the optical density changes which take place during the swelling and lysis process described by Hunter et al. (1959). It has also been reported by Hunter (1961) that 75% of the initial mitochondrial protein is found in the 100,000 x g supernatant after ascorbate treatment. Since phospholipids are known to be necessary constituents for the maintenance of mitochondrial structure it seems reasonable to think that the phospholipid alteration is the event underlying the lytic process.

With the available techniques for the separation and isolation of mitochondrial membranes we decided to go a step further and see which part of the mitochondrial structure or which structures of the mitochondrion would be affected by the ascorbate treatment.

Isolated inner mitochondrial membranes were incubated as indicated in Fig. 5. After the incubation, the suspension of inner mitochondrial membranes was chilled and centrifuged at 8500 x g for 10 min. The pellet obtained was called Fraction F-I. The supernatant was recentrifuged at 100,000 x g for one hour and another sediment was obtained; we will refer to this as Fraction F-II. The supernatant of this centrifugation will be referred to as Fraction F-III.

Each fraction was examined by electron microscopy with negative staining after osmium tetroxide fixation, following a technique similar to that of Parsons *et al.* (1965). The pellets obtained after centrifugation were resuspended in just enough unbuffered 0.25 M sucrose to give a slightly cloudy suspension. An equal volume of veronal buffered 2% osmium tetroxide, pH 7.2, was added and the mixture kept at 2°C for 30 min. Preparations for electron microscopy were obtained by placing a small drop of the suspension on Formvar-coated grids; excess liquid was removed with filter paper leaving a thin film, and then a small drop of 2% phosphotungstate, pH 7.0, was applied to the grid. Supernatant

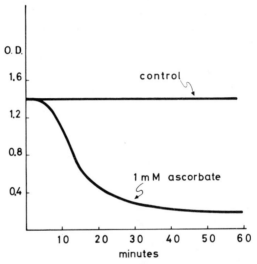

Figure 5. Optical density changes of inner mitochondrial membrane suspension during ascorbate incubation. The decrease in optical density is related to structural disaggregation. Incubation medium: 1 mM ascorbate, 20 mM Tris-HCl buffer, 0.25 M sucrose. Mitochondrial membrane protein, 13.5 mg. Temp. 30°C. Incubation volume, 30 ml.

fractions were studied by taking a small aliquot and mixing it with an equal volume of buffered osmium tetroxide and from this suspension grids were prepared as above. Electron micrographs were taken on a Siemens Elmiskop IA at 20,000 to 80,000 magnifications.

Optical density changes during the incubation of the inner mitochondrial membranes in the presence of ascorbate are shown in Fig. 5. After a 5 min lag a sharp decrease of the optical density took place reaching its lowest value around 30 to 40 min. The optical density of the controls remained completely unchanged.

Table 1 shows the values of total phospholipid P, phosphatidyl choline and phosphatidyl ethanolamine in extracts of inner mitochondrial membrane before and after incubation with and without ascorbate. Table 2 shows that the lipid P

Table 1. Total phospholipid, phosphatidyl choline and phosphatidyl ethanolamine in inner mitochondrial membranes

	Non-incubated control (μg P)	Incubated control (μg P)	+ Ascorbate (μg P)
Total phospholipid	68	67	40.5
Phosphatidyl choline	14	13	7
Phosphatidyl ethanolamine	8.7	8.5	2

Incubation medium: 1 mM ascorbate, 20 mM Tris-HCl buffer (pH 7.4), 0.25 M sucrose. Inner mitochondrial membrane protein, 13.5 mg. Final incubation volume, 30 ml. Temp. 30°C. Phospholipids were extracted after 60 min incubation.

Table 2. Effect of ascorbate on inner mitochondrial membranes (P bound to mitochondrial protein)

	Incubated Control (1) (μg P)	+ 1 mM Ascorbate (2) (μg P)	Difference between (1) and (2) (μg P)
Total phospholipid	51	39	+ 12
Protein	12	23.5	– 11.5
Dialyzed protein	12	23	– 11

Incubation medium: 1 mM ascorbate, 20 mM Tris-HCl buffer (pH 7.4), 0.25 M sucrose. Inner mitochondrial membrane protein, 9.5 mg. Final incubation volume, 10 ml. Temperature 30°C. Lipid P and P bound to protein was determined after 60 min incubation.

Table 3. Total phospholipid, protein, lipid to protein ratios, and per cent of protein recovery in different subfractions from inner mitochondrial membranes.

Fraction	Lipid P (μg)	Protein (mg)	Lipid/protein (μg/mg)	Protein recovery %
Inner membrane	68	13.5	5.03	–
Incubated control 8500 x g fraction	67	12.8	5.22	95
Fraction I	6.3	0.61	10.3	4.6
Fraction II	6.45	2.05	3.12	15.2
Fraction III	14	11	1.27	81.5

which disappeared during the incubation appeared as P bound to protein after removing exhaustively all the phospholipids. This phosphorus was strongly bound to the protein and could not be removed by dialysis against distilled water or by treating the protein with 5% trichloroacetic acid at 90°C. Table 3 gives the values of phospholipid phosphorus, protein, and lipid P to protein ratios for the inner membrane and the submembranous structures of F-I, F-II and F-III, as well as the per cent protein recovery. Protein recovered in the material sedimented at 8500 x g was as high as 95% for the incubated control, whereas it was only 4.5% for the membranes incubated with ascorbate.

Our inner membrane preparations showed the typical ultrastructure with the projecting subunits. The material sedimented at 8500 x g from incubated controls had a preserved inner membrane ultrastructure and, only very occasionally, slight structural alterations were found after incubation. An electron micrograph of this light disaggregation is shown in Fig. 6A, which has proved to be of great value for the interpretation of the material found in Fraction F-III.

Fraction F-I (Fig. 6B) is composed of smooth membranous structures completely lacking projecting subunits. Fraction F-II (Fig. 6C) is similar to Fraction F-I but its membranous structure is much thinner and contains small aggregates of projecting subunits. Fraction F-III (Fig. 6D) is very homogeneous and contains only granular structures around 80 Å in diameter with some material which tends to aggregate them.

The large decrease in optical density of the inner membrane suspension is due to structural disaggregation caused probably by the attack of ascorbate on the phospholipids.

Figure 6D appears to show areas of disruption of the inner membrane structure into particles of aggregates of projecting subunits along with their basal material. The obvious similarities of the disrupted area in Fig. 6A and the photograph of Fraction F-III (Fig. 6D) lead us to interpret it as consisting of projecting subunits along with their basal material. We also have observed that Fraction F-III contains more than 80% of the total inner membrane protein and, furthermore, the lipid P to protein ratio is much smaller than that of the whole inner mitochondrial membrane.

The smooth membranous structures of Fraction F-I (Fig. 6A) have a higher lipid P to protein ratio than that of whole inner mitochondrial membrane. This observation suggests to us that in this fraction we are dealing with a special structure of the inner membrane which has separated as such under the disruptive action of ascorbate and which in turn resisted any further lytic effect.

Fraction F-II, constituted by small fragments of submembranous structures, probably has a different nature than those of Fraction F-I, since its morphological appearance and lipid P to protein ratio are also different; however, a definite answer must wait.

The fact that phosphatidyl ethanolamine and phosphatidyl choline decrease

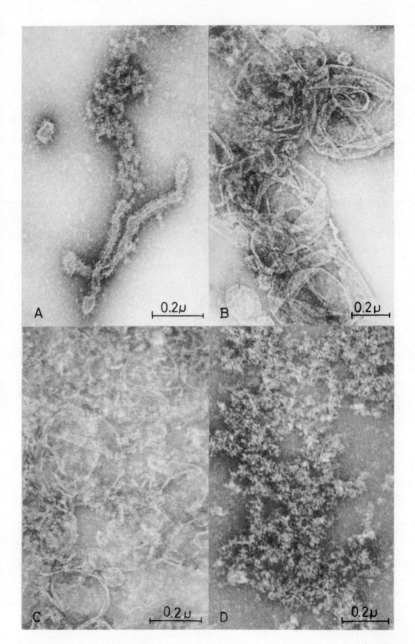

Figure 6. Electron micrographs of inner mitochondrial membrane subfractions. A: Tubular structure with projecting subunits. One of the tubules shows an area of disaggregation found very occasionally in controls incubated without ascorbate. B: Smooth membranous structures corresponding to Fraction F-I. C: Smooth membranous structures corresponding to Fraction F-II. D: Granular structures corresponding to Fraction F-III, identical in appearance to the area of disaggregation shown in A.

during ascorbate attack and the fact that different fractions do separate lead us to suggest that these phospholipids are in some way responsible for holding together different special structures of the inner mitochondrial membrane.

CYSTEINE EFFECT ON ISOLATED INNER MITOCHONDRIA MEMBRANES

Figure 7 shows that a suspension of inner mitochondrial membranes incubated in the presence of cysteine also exhibited optical density changes. However, the curve was different from those obtained with ascorbate.

Table 4. Effect of cysteine on inner mitochondrial membranes (P bound to mitochondrial protein)

	Incubated control (1) (μg P)	+ 0.8 mM Cysteine (2) (μg P)	Difference between (1) and (2) (μg P)
Total phospholipid	44.5	35.5	+ 9
Protein	11	19.4	− 8.4

Incubation medium: 0.8 mM cysteine, 20 mM Tris-HCl buffer (pH 7.4), 0.25 M sucrose. Inner mitochondrial membrane protein, 8.6 mg. Final incubation volume, 10 ml. Temperature 30°C. Lipid P and P bound to protein was determined after 120 min incubation.

Table 5. Total phospholipid, protein, lipid to protein ratios, and per cent of protein recovery in different subfractions from inner mitochondrial membranes treated with cysteine

Fraction	Lipid P (μg)	Protein (mg)	Lipid P/protein (μg/mg)	Protein recovery %
Inner membrane Incubated control	210	40	5.2	—
8500 × g fraction	200	31.5	5	80
Fraction I	131.5	11.3	11.4	30.8
Fraction II	9	1.4	6.4	6
Fraction III	1.9	23.7	0.08	61.7

Incubation medium: 0.8 mM cysteine, 20 mM Tris-HCl buffer (pH 7.4), 0.25 M sucrose. Inner mitochondrial membrane protein, 40 mg. Final incubation volume, 30 ml. Temperature 30°C. Phospholipids were extracted after 120 min incubation.

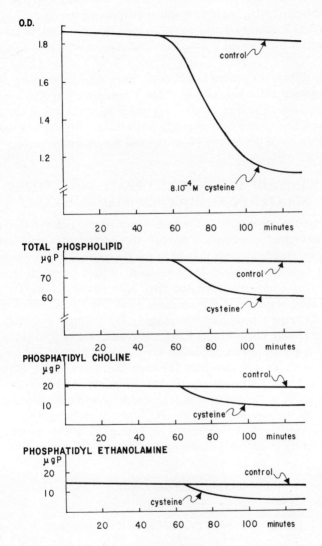

Figure 7. Optical density, total phospholipid, phosphatidyl choline and phosphatidyl ethanolamine changes of a suspension of inner mitochondrial membranes incubated in the presence of 0.8 mM cysteine. Phospholipids were determined in aliquots after 20, 40, 65, 80, 100 and 120 min.

It can be seen that the decrease in optical density coincided with a progressive decrease in phospholipids. The lipid P which disappeared during the incubation appeared again as P strongly bound to the protein fraction. Table 4 shows that the lipid P which disappeared during the incubation appeared as P bound to protein.

After incubation in the presence of cysteine, three subfractions were obtained through differential centrifugation. Fraction I sedimented at 8500 x g. The supernatant was centrifuged at 100,000 x g and Fraction II was obtained. The resulting supernatant of this centrifugation was Fraction III. Table 5 shows the lipid to protein ratio of the different fractions and the protein recovery.

MECHANISM OF ACTION OF ASCORBATE AND CYSTEINE ON ISOLATED INNER MITOCHONDRIAL MEMBRANES

Most probably the mechanism of action of ascorbate and cysteine on the isolated mitochondrial membrane is through the formation of lipid peroxides from the unsaturated fatty acids present in the phospholipids of the membrane. Cysteine and ascorbate catalyze the formation of lipid peroxides during the incubation of emulsions of unsaturated fatty acids (Wilbur et al., 1949). The formation of lipid peroxides in mitochondria incubated with ascorbate has been observed in different laboratories (Ottolenghi, 1959; Thiele and Huff, 1960; Hunter et al., 1964a). On the other hand, the production of lipid peroxides in mitochondria is greatly enhanced during the auto-oxidation of reducing compounds such as thiols and ascorbate (Fortney and Lynn, 1964; Skrede and Christophersen, 1968; Hunter et al., 1964b). Wills (1965) has considered that the peroxides could be formed from unsaturated fatty acids; this reaction would be catalyzed by hemoproteins, or by Fe^{3+} plus a reducing agent such as ascorbate or cysteine. Since thyroxine is known to inhibit mitochondrial swelling induced by glutathion (Hunter et al., 1964c), Fe^{2+} and ascorbate (Cash

Table 6. Effect of thyroxine on phospholipids of inner mitochondrial membranes incubated with cysteine

	Non-incubated control (μg P)	Incubated control (μg P)	+ 0.8 mM Cysteine (μg P)	+ 10^{-5} M Thyroxine (μg P)	+ 0.8 mM Cysteine +10^{-5} M thyroxine (μg P)
Total phospholipid	74	66	53	70	68
Phosphatidyl choline	20	19	13	20.1	20.3
Phosphatidyl ethanolamine	18	18.2	10	18	18.1

Incubation medium: 20 mM Tris-HCl buffer (pH 7.4), 0.25 M sucrose. Inner mitochondrial membrane protein, 12.5 mg. Final incubation volume, 10 ml. Temperature 30°C. Phospholipids were extracted after 2 h incubation.

et al., 1966) through an anti-oxidant action, we decided to study the effect of thyroxine on the lysis of the inner membrane induced by cysteine or ascorbate.

We have found that concentrations of thyroxine (10^{-5} M) completely inhibit the decrease in optical density of a suspension of mitochondrial membranes incubated in the presence of ascorbate or in the presence of cysteine. This

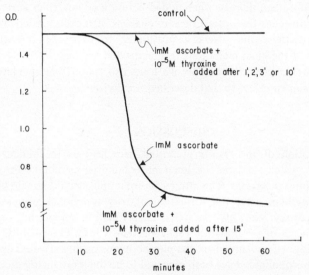

Figure 8. Effect of thyroxine on optical density changes at 520 mμ of a suspension of inner mitochondrial membranes incubated in the presence of ascorbate. Thyroxine was added to different tubes at the indicated times of incubation. Final thyroxine concentration, 10^{-5} M.

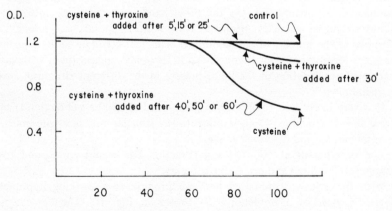

Figure 9. Effect of thyroxine on optical density changes at 520 mμ of a suspension of inner mitochondrial membranes incubated in the presence of cysteine. Thyroxine was added to different tubes at the indicated times of incubation. Final thyroxine concentration, 10^{-5} M.

concentration of thyroxine also inhibits the decrease in phospholipid content (see Table 6).

In a different set of experiments we have studied this inhibitory effect of thyroxine when added at different times during the incubation in the presence of ascorbate or cysteine. It can be seen in Figs. 8 and 9 that thyroxine inhibits the decrease in optical density when added before a certain time during the lag period which precedes the optical density changes of the suspension. We think, therefore, that thyroxine offers a possibility to carry out the structural disaggregation of the inner membrane in a controlled manner.

In some preliminary experiments we have also found that tocopherol and EDTA have a similar effect to that described for thyroxine.

DISCUSSION

The interpretation of the results presented here lead us to think that both ascorbate and cysteine act on the inner mitochondrial membrane through the formation of lipid peroxides. The altered phospholipids would remain bound to the protein, perhaps after the formation of new functional groups such as carboxylic, hydroxy, oxo, etc., as it has been found in compounds originated during the process of peroxidation (Schauenstein, 1967). Phosphatidyl choline and phosphatidyl ethanolamine are the phospholipids mainly affected during the lysis. These phospholipids have been reported to be the more highly unsaturated in the mitochondria (Huet et al., 1968) and therefore they would be the main source for the formation of peroxides.

After reaching a certain degree of phospholipid alteration the membrane would disaggregate, therefore removing the weak linkages between the lipids and proteins of the membrane. On the other hand, our structural studies of the submitochondrial particles after the lysis, seem to point to a selective disassembly of the mitochondrial membrane. This fact could be easily explained if the inner membrane is not homogenous in its lipidic composition, with varying degrees of unsaturation of the lipids present in the different structural areas in the membrane.

It is very likely that the fractions sedimented at 8500 x g (Fraction I) and at 100,000 x g (Fraction II) correspond to the lamellar structures present in the inner membrane and described above.

The supernatant at 100,000 x g (Fraction III) would correspond to the disaggregated tubules and projecting subunits.

With the evidence presented here we propose as a working hypothesis a schematic view of a mitochondrion shown in Fig. 10. The "cristae mitochondriales" would correspond to the described lamellar structures and the tubules covered by projecting subunits would be as branches inserted on the cristae.

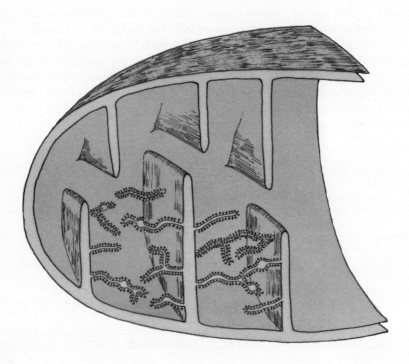

Figure 10. Diagrammatic representation of mitochondrion, showing the outer membrane, the cristae and the tubules covered with projection subunits.

SUMMARY

Studies carried out in our laboratory have shown that rat liver inner mitochondrial membranes examined by electron microscopy after negative staining exhibit two different types of structures; i.e. lammellae and tubules. The tubules have the typical projecting subunits covering their entire outer surface, while the lamellae, representing a relatively minor component, completely lack the projecting subunits. When suspensions of isolated inner mitochondrial membranes were incubated at 30°C in 0.25 M sucrose, 20 mM Tris-HCl buffer (pH 7.4), in the presence of 1 mM ascorbate or 0.8 mM cysteine, a marked decrease of the optical density at 520 mμ was observed. These optical density changes seem to be associated with the lysis of the inner mitochondrial membrane, which in turn depends on alterations of its phospholipid composition, consisting mainly of a decrease in phosphatidyl ethanolamine and phosphatidyl choline. The amount of lipid P which disappeared during the incubation was found to be strongly bound to the lipid extracted protein residue. Both optical density changes and phospholipid alterations were inhibited by 10^{-5} M thyroxine. Differential centrifugation of the disrupted inner membranes after one hour incubation in the presence of ascorbate has led to the separation of several types of submitochondrial particles, which were different not only in morphological appearance but also in chemical composition. These results seem to indicate that a selective disassembly of structures present in the inner mitochondrial membrane has been achieved.

REFERENCES

Cash, W. D., Gardy, M., Carlson, H. E. and Ekong, E. A. (1966). *J. biol. Chem* **241**, 1745.
Fernandez-Morán, H. (1962). *Circulation* **26**, 1039.
Fernandez-Morán, H., Oda, T., Blair, P. V. and Green, D. E. (1964). *J. Cell Biol.* **22**, 63.
Fortney, S. R. and Lynn Jr., W. S. (1964). *Archs Biochem. Biophys.* **104**, 241.
Huet, C., Levy, M. and Pascaud, M. (1968). *Biochim. biophys. Acta* **150**, 521.
Hunter, Jr., F. E. (1961). In "Biological Structure and Function", (T. W. Goodwin and O. Lindberg, eds.), Vol. 2, p. 53, Academic Press, London and New York.
Hunter, Jr., F. E., Levy, J. F., Fink, J., Schutz, B., Guerra, F. and Hurwitz, A. (1959). *J. biol. Chem.* **234**, 2176.
Hunter, Jr., F. E., Scott, A., Hoffsten, P. E., Guerra, F., Weinstein, J., Schneider, A., Schutz, B., Fink, J., Ford, L. and Smith, E. (1964a). *J. biol. Chem.* **239**, 604.
Hunter, Jr., F. E., Scott, A., Hoffsten, P. E., Gebicki, J. M., Weinstein, J. and Schneider, A. (1964b). *J. biol. Chem.* **239**, 614.
Hunter, Jr., F. E., Scott, A., Weinstein, J. and Schneider, A. (1964c). *J. biol. Chem.* **239**, 622.
Mitchell, R. F. (1967a). *J. Ultrastruct. Res.* **18**, 257.
Mitchell, R. F. (1967b). *J. Ultrastruct. Res.* **18**, 277.
Ottolenghi, A. (1959). *Archs Biochem. Biophys.* **79**, 355.
Parsons, D. F. (1963). *Science, N. Y.* **140**, 985.
Parsons, D. F., Williams, G. R. and Chance, B. (1965). *Ann, N.Y. Acad. Sci.* **137**, 643.
Santiago, E., Vázquez, J., Guerra, F. and Macarulla, J. M. (1968a). *Revta esp. Fisiol.* **24**, 1931.
Santiago, E., Guerra, F. and Macarulla, J. M. (1968b). *Revta esp. Fisiol.* **24**, 25.
Schauenstein, E. (1967). *J. Lipid Res.* **8**, 417.
Sjöstrand, F. S., Anderson-Cedegren, E. and Karlsson, U. (1964). *Nature, Lond.* **202**, 1075.
Skrede, S. and Christophersen, B. O. (1968). *Biochem. J.* **106**, 515.
Stoeckenius, W. (1963). *J. Cell Biol.* **17**, 443.
Thiele, E. H. and Huff, J. W. (1960). *Archs Biochem. Biophys.* **88**, 203.
Wilbur, K. M., Bernheim, F. and Shapiro, O. W. (1949). *Archs Biochem. Biophys.* **24**, 305.
Wills, E. D. (1965). *Biochim. biophys. Acta* **98**, 238.

Molecular Anatomy of a Membrane

J. S. O'BRIEN

Department of Neurosciences, School of Medicine, University of California at San Diego, La Jolla, California, U.S.A.

The world of membranes is no longer simple. Membranes vary in composition, structure and function from species to species, from cell to cell and from one intracellular locale to another. They are dynamic, plastic, fluid structures which have no fixed state. We fall into the trap of naïveté when we describe the molecular anatomy of membranes with simple molecular models. Such models fall short of reality; those who believe that they are divinely revealed are misled. Nonetheless, molecular models are useful in understanding the nature of membranes since they provide a framework upon which to shape, reshape and embellish our thoughts about these interesting biological structures.

I have turned my attention to the study of the matriarch of membranes, the myelin membrane, in an attempt to comprehend, in some small measure, the reasons for its normal stability and for its instability in disease states. Myelin has the following attributes: (i) it is the most stable membrane known, (ii) it can be easily prepared pure in good yield, (iii) it is a lipid-rich membrane (80% lipid, 20% protein), and (iv) the lipid composition and the fatty acid composition of each lipid are known (O'Brien and Sampson, 1965a,b).

There are approximately 1500 different lipid molecules in human central nervous system myelin, but only about thirty predominate. Molecular models of each common lipid were made and these were assembled in bimolecular leaflet fashion, packing them as closely as possible, setting the distances to conform to those determined by X-ray diffraction analyses. A similar model was constructed for the brain mitochondrial membrane, from data on the lipid composition and the fatty acid composition of each lipid in this organelle (O'Brien, 1967).

Rather striking differences were apparent when the myelin and the mitochondrial models were compared (Figs. 1 and 2). Myelin contained more cholesterol, more long chain sphingolipids and fewer polyunsaturated lipids than mitochondria. It was possible to pack myelin lipids much closer than mitochondrial lipids. This was chiefly due to the presence of many more polyunsaturated lipids in mitochondria. These lipids are much bulkier than saturated lipids; the introduction of *cis* double bonds in their fatty acyl chains causes the chains to be bent into hook-like or snake-like configurations.

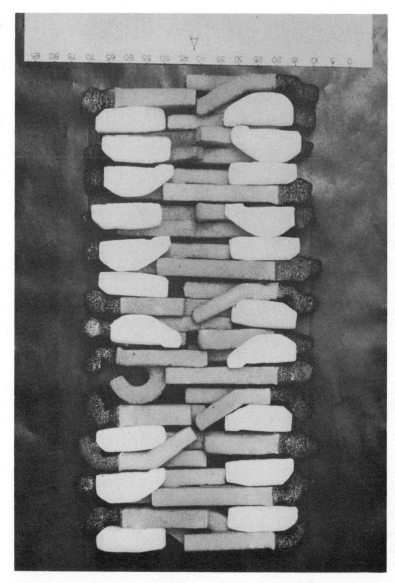

Figure 1. Molecular model of myelin lipid bilayer. Molecular models of myelin lipids were constructed from Dreiding stereo-models in styrofoam and assembled with their polar head groups 55 Å-60 Å apart. Cholesterol molecules are white.

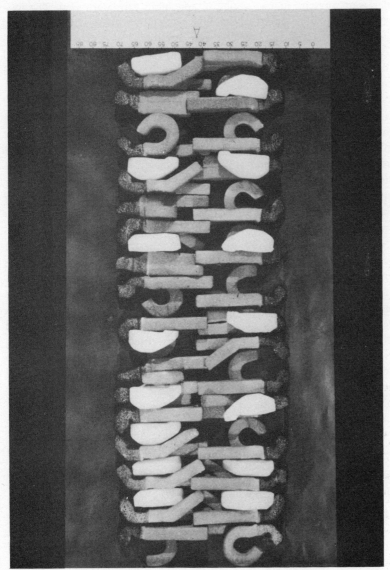

Figure 2. Molecular model of mitochondrial membrane lipids in bilayer configuration.

Cholesterol filled in many of the gaps between the lipid head groups in myelin and enhanced intermolecular cohesion in this way. Fewer cholesterol molecules were present in mitochondria and more gaps were seen. The long chain sphingolipids also enhanced stability by sending their fatty acyl chains across the center of the lipid bilayer and interdigitating with the fatty acyl chains of the molecules on the opposite side of the membrane. Myelin contained 5-fold higher proportions of these long chain sphingolipids than mitochondria (O'Brien, 1967).

I have proposed (O'Brien, 1965) that long chain sphingolipids are of special importance in the cohesional stability of myelin for the following reasons: (i) the highest concentrations of long chain sphingolipids occur in myelin, the most stable membrane known, (ii) physical studies of these molecules indicate that they are bilayer stabilizers, and (iii) the proportions of these molecules are greatly reduced in some diseases in which myelin is unstable. Recent experiments have supported the hypothesis that the acyl groups of these lipids are the major determinants of their stabilizing effect; the longer the chain length of the fatty acyl group, the slower the turnover of the myelin lipid (O'Brien, 1967).

The striking differences in the configuration of the myelin model and the mitochondrial model emphasize the fallacy inherent in drawing conclusions about membranes in general from considerations of only a few specific types. Both models fail to include proteins known to be important membrane constituents. In the case of myelin this omission is not so grave; it is difficult to conceive of proteins playing a major role, since their proportions are low and since myelin lipids tend to spontaneously form stable bilayers. In myelin, proteins are probably present in the extended β-keratin configuration apposed to the lipid polar groups. In the case of mitochondria, the omission of protein is more important; proteins are major constituents and the lipids do not form stable bilayers by themselves. In mitochondria, proteins are probably present in coiled configurations, with the acyl chains of lipid molecules inserted into the protein interacting with stretches of hydrophobic amino acids, as Professor Benson has suggested (Benson, 1966). These lipoprotein subunits are held together by hydrophobic associations resulting from an increase in entropy derived from transferring hydrophobic groups from an aqueous to a non-aqueous phase. Thus, the integrity of myelin can be thought of as lipid-derived, whereas the integrity of mitochondria can be thought of as protein-derived.

The tightly organized myelin-type lipid bilayer and the somewhat more loosely organized mitochondrial globular lipoprotein subunit can be thought of as the extremes in the molecular anatomy of membranes. It is probable that most membranes will fall between these extremes. It is useful to think of the functions of these two types of membranes. Membranes like myelin, such as plasma membranes, act as permeability barriers to the passage of water soluble molecules. The condensed bilayer configuration is ideally suited to the role of permeability barrier to the indiscriminate passage of water soluble molecules. Membranes like mitochondria act as surfaces upon which solid state reactions are carried out. They need not be concerned with permeability barrier functions since they are enclosed by external membranes which perform for them the service of molecular screening. Their molecular anatomy is less rigid and they are comprised of subunits so that conformational changes, oscillations, peristalsis, and contractions are not restricted. The hydrophobic interior of the organelle membrane acts to provide a medium of low dielectric constant so that electron transfer reactions and hydrolytic reactions are spacially oriented towards the surface of the membrane.

As a final thought, it is becoming apparent that human diseases can result from molecular alterations of membrane structure. We have studied two demyelinating diseases in humans which apparently are due to chemically altered myelin. Metachromatic leucodystrophy is a demyelinating disease transmitted as an autosomal recessive trait. In this disease myelin contains a 4-fold excess of cerebroside sulfate compared to normal (O'Brien and Sampson, 1965c). Demyelination in this disease may be a result of the abnormality in the chemical composition of the membrane, since (i) the excess number of sulfate groups leads to an abnormally electronegative surface charge with resultant electrostatic repulsion of juxtaposed membranes, (ii) conformational changes in the structure of the myelin protein may also result, and (iii) the sheath is a less effective ionic insulator due to its increased affinity for cations.

Refsum's disease is another demyelinating disease transmitted as an autosomal recessive trait. An accumulation of phytanic acid, 4,7,11,15-tetramethyl hexadecanoic acid, occurs throughout the body and, as we have shown (MacBrinn and O'Brien, 1968), also in myelin, particularly in the 1-position of lecithin. Demyelination in this disease may be due to disruption of the packing of the hydrocarbon-tail region of the membrane due to steric hindrance imposed by the branched methyl groups of phytanoyl-containing lecithin with resultant membrane destabilization. In support of this conjecture is the demonstration that nerve function in Refsum's disease improves on a phytanic acid-free diet.

These diseases are among the first to be described in man which can be classified as membranopathies. As we achieve a clear understanding of the molecular basis of such disorders, rational approaches to therapy will be possible.

REFERENCES

Benson, A. A. (1966). On the orientation of lipids in chloroplast and cell membranes. *J. Am. Oil Chem. Soc.* **43**, 265.

MacBrinn, M. C. and O'Brien, J. S. (1968). Lipid composition of the nervous system in Refsum's disease. *J. Lipid Res.* **9**, 552.

O'Brien, J. S. (1965). The stability of the myelin membrane. *Science, N.Y.* **147**, 1099.

O'Brien, J. S. (1967). Cell membranes: composition, structure, function. *J. Theoret. Biol.* **15**, 307.
O'Brien, J. S. and Sampson, E. L. (1965a). Lipid composition of the normal human brain: gray matter, white matter and myelin. *J. Lipid Res.* **6**, 537.
O'Brien, J. S. and Sampson, E. L. (1965b). Fatty acid and fatty aldehyde compositions of the major brain lipids in normal human gray matter, white matter and myelin. *J. Lipid Res.* **6**, 545.
O'Brien, J. S. and Sampson, E. L. (1965c). Myelin membrane: a molecular abnormality. *Science, N.Y.* **150**, 1613.

Properties of the Purified Cytoplasmic Membrane of Yeast

Ph. MATILE

Department of General Botany, Swiss Federal Institute of Technology, Zurich, Switzerland

1. INTRODUCTION

Elucidation of the structure and function of a certain type of membrane must be based on the detailed knowledge about its chemical composition. Therefore, the isolation and purification of the membrane is an indispensable prerequisite of the respective investigations. In the case of the plasma membrane of yeast cells several authors have reported recently on its isolation procedures as well as its chemical composition. Although the results presented so far do not permit the establishment of well-defined structural and functional roles of certain components identified, they nevertheless point to some unique properties of the yeast plasmalemma.

2. ISOLATION OF THE YEAST PLASMALEMMA

One method for isolating plasma membranes is based on the osmotic bursting of yeast sphaeroplasts (Boulton, 1965; Garcia Mendoza and Villanueva, 1967; Longley et al., 1968). In the presence of Mg^{2+} the membranes are stable and can be isolated by centrifugation. However, one shortcoming of this method concerns the possible degradation of components of the plasmalemma by the action of digestive enzymes used for preparing sphaeroplasts. In addition, the use of Mg^{2+} for stabilizing the membranes may result in the formation of aggregates between plasmalemma and other components of the lysed sphaeroplasts.

Suomalainen et al. (1967) prevent this possible contamination by first isolating cell walls which apparently still contain associated plasmalemma; the subsequent digestion of the cell walls by snail gut juice yields membranes resembling the "ghosts which are seen in protoplast lysates". This method seems to be characterized by considerable problems concerning the complete removal of the cell wall.

A method developed recently (Matile et al., 1967) is based on the fractionation of extracts from whole yeast cells using differential and density gradient centrifugation. The most important step of this technique is the isopycnic centrifugation in gradients of Urografin. Using freeze-etching techniques membranes equilibrating at ca. 1.17 g/cm^{-3} could be identified as fragments of the plasmalemma.

This method has now been modified because initially its application was limited to membranes from cells which were cultivated anaerobically in the absence of ergosterol and unsaturated fatty acids. Membranes from aerobically grown cells appeared to be much less stable in the presence of the medium used (sucrose, Tris-HCl buffer, EDTA) than are plasma membranes from anaerobically grown cells. The stability of the former could be achieved, however, by using an unbuffered sucrose medium (10% w/v). The cells (*S.cerevisiae*) were ruptured by vigorous shaking in the presence of ballotini beads and a cell-free extract was obtained by centrifugation for 20 min at 1500 g. The extract (100 ml from 10 ml of packed cells) was then layered on discontinuous gradients with steps of 55, 40 and 30% (w/v) sucrose and centrifuged for 3 h in an angle head rotor at 48,000 g. The plasmalemma, contaminated with glycogen and some mitochondrial material, was present in the sediment which was resuspended in sucrose medium containing 10 mg/ml of pancreatic amylase. After 2 h incubation at 4°C the membranes were collected by centrifugation (15 min, 150,000 g.) The resulting crude plasmalemma fraction was subsequently centrifuged in density gradients of Urografin (methylglucamine salt of N,N'-diacetyl-3,5-diamino-2,4,6-triiodobenzoic acid) ranging from 1.11 to 1.25 g/cm^{-3} (Spinco rotor SW 25; 5 h, 25,000 rpm). The particles present in the strong band, whose position corresponded to a density of Urografin of ca. 1.175 g/cm^{-3}, were collected from several gradients and washed three times with distilled water (centrifugations: 15 min, 150,000 g). The resulting purified plasmalemma formed an almost transparent, gelatinous pellet.

3. FINE STRUCTURE OF THE YEAST PLASMALEMMA

In frozen-etched specimens the plasmalemma of yeast cells shows a particulate structure on the outer surface. The globules have a diameter of ca. 150 Å and are embedded in the membrane at a variable depth (Moor and Mühlethaler, 1963). This observation may suggest that the particles are moving through the plasmalemma. It is interesting to note that particles having a similar dimension sculpture the fracture planes across the cell walls (Fig. 1A). Evidence has been presented that the plasmalemma particles are mannanprotein in nature (Matile *et al.*, 1967). Thus, it may be speculated that they represent building stones of the cell wall.

The purity of plasma membranes, prepared as described above (Fig. 1B), and the identity of the vesicles, is demonstrated by the freeze-etching which shows the characteristic particulate structure on their surface (Fig. 1C). Frequently, the fracturing of the frozen specimens seems to result in a partial detachment of the

PROPERTIES OF THE PURIFIED CYTOPLASMIC MEMBRANE OF YEAST 41

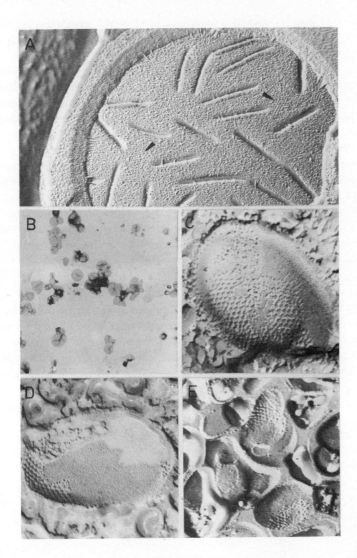

Figure 1. (A) View on the outer surface of the plasmalemma of *Saccharomyces cerevisiae*. The freeze etching shows particulate structures occasionally concentrated in hexagonal arrays (arrows). The invaginations are free of particles. Note the particles which sculpture the fracture plane across the cell wall. (Magnification × 44,000). (B) Isolated fragments of the plasmalemma placed on a formvar film and fixed with OsO_4 vapour. (Magnification × 5600). (C, D) Isolated plasmalemma. Note the hexagonally arranged particles. In (D) the area of the membrane which is free of particles is characterized by a fine granular structure. (Magnification: (C) × 84,800; (D) × 80,000). (E) Isolated plasmalemma extracted with 0.5% deoxycholate. (Magnification × 64,000).

globules and possibly also of a superficial layer of the membrane. As shown in Fig. 1D, the respective areas possess a fine granular structure (structural protein?).

Thin sections of pellets fixed with glutaraldehyde and post-fixed with osmium tetroxide show vesicles having a triple-layered structure (Fig. 2B). Upon staining of the section with lead citrate, electron dense deposits are associated with the membranes (Fig. 2A). This reaction with lead ions could be due to the presence of mannanproteins containing phosphate groups (see §6). Isolated plasmalemma negatively contrasted with phosphotungstate does not show any remarkable fine structure (Fig. 2C). It is interesting to note that the particulate surface structure visible on freeze-etchings can be seen neither with the negative contrast technique nor in specimens subjected to shadow casting.

The vesicles of isolated plasmalemma have diameters ranging from 0.2 to 0.7 μ. They are much smaller than the ghosts of sphaeroplasts and, therefore, represent fragments of cytoplasmic membranes.

4. CHEMICAL COMPOSITION OF ISOLATED YEAST PLASMA MEMBRANES

Data on the composition of isolated yeast plasma membranes reported by several authors are compiled in Table 1. They strongly suggest the lipoprotein nature of this membrane. A comparison of the overall compositions of membranes isolated with the methods described above shows considerable differences, however. The most important qualitative difference concerns the nucleic acids which are

Table 1. Composition of plasma membranes isolated from sphaeroplasts, cell walls or whole yeast cells (the figures express percentages of dry membrane preparations)

Authors	Boulton (1965)	Garcia-Mendoza and Villanueva (1967)	Longley et al. (1968)	Suomalainen et al. (1967)	Matile (unpublish. data)
Object	S. cerevisiae	Candida utilis	S. cerevisiae	S. cerevisiae	S. cerevisi
Protein	46-47.5	38.5	49.3	ca. 30-40	26.6
RNA	6.7	1.1	7.0	–	0
DNA	0.97	0	–	–	0
Carbohydrate	3.2	5.2	4.0-6.0	ca. 25	30.8
Lipid	37.8-45.6	40.4	39.1	ca. 30-40	45.5
Sterols (ergosterol)	ca. 5.6	–	6.0	–	8.1
Total phosphorus	1.08	–	1.21	–	0.86
Lipid phosphorus	ca. 0.4	–	0.25	–	0.186
Phospholipid (lecithin)	ca. 9.5	–	ca. 5.7	–	4.24

PROPERTIES OF THE PURIFIED CYTOPLASMIC MEMBRANE OF YEAST

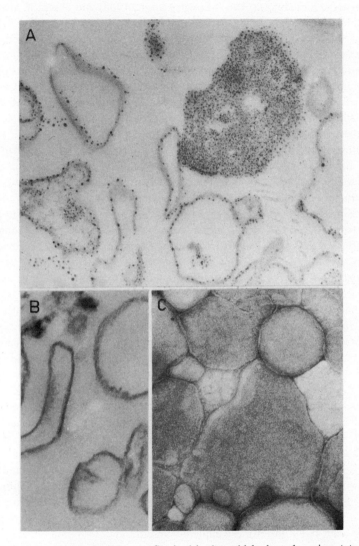

Figure 2. (A) Isolated plasmalemma fixed with glutaraldehyde and osmium tetroxide, stained with uranylacetate and lead citrate. (Magnification x 57,600). (B) Isolated plasmalemma fixed with osmium tetroxide. Note triple-layered membrane structure. (Magnification x 57,600). (C) Isolated plasmalemma negatively contrasted with phosphotungstate. (Magnification x 56,000).

present in the preparations obtained from sphaeroplasts (Boulton, 1965; Garcia Mendoza and Villanueva, 1967; Longley et al., 1968) and absent in membranes obtained from whole cells (Matile, unpublished data). It seems that the Mg^{2+} causes unspecific aggregations between nucleic acids and components of the plasmalemma. A conspicuous quantitative difference concerns the amount of carbohydrate, which is much lower in the ghosts from sphaeroplasts (3-6%) than in the membrane fragments from whole cells (ca. 30%). Acid hydrolysis of the purified defatted plasmalemma yields mannose only. It seems that the high mannan content is responsible for the comparatively high equilibrium density of the plasma membrane in gradients of Urografin (1.17-1.18 g/cm^{-3}); if the membranes from lysed sphaeroplasts are centrifuged in the same gradient system they equilibrate at densities around 1.10 g/cm^{-3}. The low amount of polysaccharide associated with the ghosts of sphaeroplasts is possibly due to a partial digestion or release of this material during sphaeroplast formation.

The percentage of lipids seems to be relatively independent of the method used for isolating the membranes. The values reported fall into a range from 35-45%. Longley et al. (1968) have presented a detailed analysis of sphaeroplast ghosts showing the presence of various mono-, di-, and tri-glycerides, sterols and phospholipids. We have used thin layer chromatography (silica gel; solvent: butylacetate) for separating sterols from other lipids. A strongly UV-absorbing spot had an r_F value of 0.55-0.58 and a UV spectrum identical with that of ergosterol. Using more effective methods, Longley et al. (1968) have been able to resolve the sterol fraction into several components. These membrane constituents seem to be responsible for the susceptibility of yeast cells and sphaeroplasts to polyene antibiotics which are known to interact with sterols (cf. Lampen, 1966). In fact, isolated plasmalemma binds fungicides like amphotericin and nystatin; as a result the membranes are partially disintegrated (Matile, 1970). If the silica gel layers are sprayed with anisaldehyde-sulfuric acid (Stahl, 1962), a reagent for sterols and sugars, spots with r_F values of 0.13, 0.3, 0.48 and 0.75 appear in addition to the sterol spot. The most prominent spot (r_F, 0.3) is violet; the respective compound does not absorb UV and reacts positively with ninhydrin. It is possibly identical with one of the sphingolipids isolated from *Candida utilis* (Wagner and Zofcsik, 1966). It is interesting to note that in lipid extracts from purified cell walls the same possible glycolipids are also present, but ergosterol is completely absent.

5. ENZYMES OF THE PURIFIED PLASMALEMMA

In purified preparations of plasmalemma isolated from whole cells, various enzymes known to be localized in the ground cytoplasm, in vacuoles or mitochondria are absent. In contrast, in preparations of ghosts from lysed sphaeroplasts the presence of oxidoreductases (contaminating mitochondria ?)

has been demonstrated (Boulton, 1965). The prominent enzyme activity associated with the purified plasmalemma is that of an Mg^{2+}-dependent ATPase; this enzyme differs from the mitochondrial ATPases in being oligomycin insensitive (Matile *et al.*, 1967). A comparatively low specific activity of invertase has been found occasionally in freshly prepared membranes (Matile *et al.*, 1967). However, subsequent washings with a buffered sucrose medium (Tris-HCl, EDTA) resulted in the complete loss of this enzyme. If an unbuffered sucrose medium is used, invertase seems to be firmly attached to the plasmalemma. In density gradients of Urografin the distribution of invertase corresponds exactly with that of ATPase and mannan (Fig. 3). If compared with the

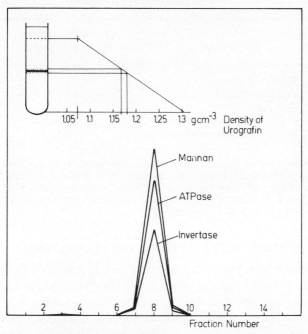

Figure 3. Purified plasmalemma centrifuged in a density gradient of Urografin. The gradient was loaded with particles suspended in 10% sucrose and spun for 2.5 h at 39,000 rpm (Spinco rotor SW 39). Distribution of invertase, ATPase and mannan.

high specific invertase activity associated with purified cell walls, the activity localized in the plasmalemma appears to be very low (Table 2). These results are in agreement with Lampen's (1968) report on the localization of invertase in the sphaeroplast membrane. Yeast invertase is known to occur in a large external and a small internal form: the former is a mannanprotein which is integrated into the cell wall; the latter is a protein localized in the cytoplasm (Gascón and Ottolenghi, 1967; Lampen *et al.*, 1967; Lampen, 1968). It has been assumed that the internal invertase represents a precursor molecule of the external invertase (Gascón and Lampen, 1968). Sphaeroplasts secrete the glycoprotein type of invertase exclusively, and it may therefore be speculated that the plasma membrane is the seat of reactions which result in the conversion of the internal into the external invertase. Preliminary studies indicate that the molecular sieving properties (Sagavac 8C) of a major fraction of the plasmalemma-bound invertase correspond with those of the external invertase, but small fractions of invertase seem to have both higher and lower molecular sizes than the external enzyme.

Table 2. Association of invertase activity with purified plasma membranes of *Saccharomyces cerevisiae*

Preparation	Invertase activity per unit of protein
Ruptured cells	73.9
Cell free extract	47.2
Cell walls	165.5
Crude plasma membrane	0.685
Purified plasma membrane	1.17

Cell walls were prepared according to the method of Garcia-Mendoza and Villanueva (1963), which was modified by treating the isolated walls with 0.5% Triton-X-100; this treatment, followed by three washings with distilled water, eliminated all of the ATPase activity which had indicated an appreciable amount of contaminating plasmalemma. Invertase activity is expressed in arbitrary units.

6. SUBFRACTIONATION OF THE PLASMALEMMA

In a previous report (Matile et al., 1967) the disintegration of isolated plasmalemma in the presence of deoxycholate (DOC) has been described. In the case of membranes obtained from anaerobically grown cells, DOC causes an extensive disintegration. The material sedimentable after DOC treatment contains large particles which can be isolated using a Urografin gradient system. They contain mannan and protein and are probably identical with the plasmalemma globules discovered by Moor and Mühlethaler (1963).

Although the morphological examination of plasma membranes from aerobically grown cells shows the presence of these particles, we have not been able to isolate them from preparations treated with DOC. Upon centrifugation of the DOC-insoluble material in density gradients of Urografin, a faint band is formed in the region which corresponds to the density of plasmalemma particles. Almost all of the material is concentrated in a band with a slightly lower density (1.16 g/cm^{-3}) than the intact plasmalemma (1.175 g/cm^{-3}). However, the composition of this material differs markedly from that of the intact plasmalemma.

As shown in Table 3, practically all of the invertase activity, a major fraction of the ATPase activity, about one third of the protein, and more than one half of the mannan is solubilized upon DOC treatment. The freeze-etchings of the DOC-insoluble material suggest that the large mannanprotein particles are still attached to membrane vesicles which, however, appear to be much smaller than those of the untreated preparations (Fig. 1E). The polysaccharide solubilized upon treatment with DOC has been precipitated in the presence of 50% methanol. About 80% of the precipitated material is water soluble and consists of 86% mannan, 13% protein and 1% phosphate. If filtered through a column of Sephadex G-200 it is eluted in a uniform peak shortly behind the front and, therefore, has a molecular size similar to that of the external invertase.

The results summarized above suggest that several different forms of mannanprotein are present in the yeast plasmalemma.

Table 3. Effect of treating purified plasmalemma of *Saccharomyces cerevisiae* with 0.5% deoxycholate.

	Purified plasmalemma	DOC-treated plasmalemma	% solubilized by DOC treatment
Protein (μg)	485	320	34.0
Mannan (μg)	1090	457	58.0
Invertase	1.01	0.036	96.4
ATPase	2.91	0.61	79.0

A pellet of membranes was resuspended in the detergent (pH 7.0) and the particulate material was subsequently collected by centrifugation for 30 min at 150,000 g. The activities of ATPase and invertase are given in arbitrary units.

7. DISCUSSION

Several components associated with the isolated plasmalemma seem to be related with components of the cell wall. It may be speculated, therefore, that one of the functions of the plasma membrane concerns the secretion of certain building stones of the cell wall. This assumption is supported by several observations: sphaeroplasts are known to secrete several external enzymes, mannan and other cell wall constituents (cf. Lampen, 1968). In addition, a close serological relationship between protoplast membranes and cell walls has been detected by Garcia Mendoza *et al.* (1968). Finally, preliminary evidence for the localization of glycolipids in both plasmalemma and cell wall of *Saccharomyces cerevisiae* has been obtained.

However, Eddy and Longton (1969) have been able to obtain a large molecular weight mannanprotein from isolated yeast cell walls treated with snail gut juice whose properties differed from that of a mannanprotein obtained from cytoplasmic membranes. Therefore, these authors have doubts about a simple translocation function of the plasmalemma. The example of invertase suggests, however, that extensive chemical changes taking place in the plasmalemma are responsible for the differences between external and membrane-bound forms of mannanproteins. In addition, it is possible that the plasmalemma is involved not only in the transfer of cell wall constituents from the cytoplasm into the periplasmatic space but also in their synthesis. Algaranti *et al.* (1963) and Behrens and Cabib (1968) have used membrane isolates from yeast cells for *in vitro* synthesis of mannan; furthermore, it has been reported that the biosynthesis of sphingolipids is catalyzed by microsomal preparations from yeast cells (Braun and Snell, 1967, 1968; Stoffel *et al.*, 1968). It will be a future task to examine whether these activities are associated with the isolated plasmalemma.

ACKNOWLEDGEMENTS

The present work has been supported by a grant of the Swiss National Science Foundation and by the Schering AG, Berlin (supply of a large amount of Urografin). The assistance of Miss R. Rickenbacher, W. Guyer and F. Kopp is gratefully acknowledged.

REFERENCES

Algaranti, I. D., Caminatti, H. and Cabib, E. (1963). *Biochem. biophys. Res. Commun.* **12**, 504.
Behrens, N. H. and Cabib, E. (1968). *J. biol. Chem.* **243**, 502.
Boulton, A. A. (1965). *Expl Cell Res.* **37**, 343.
Braun, P. E. and Snell, E. E. (1967). *Proc. natn. Acad. Sci. U.S.A.* **58**, 298.
Braun, P. E. and Snell, E. E. (1968). *J. biol. Chem.* **243**, 3775.
Eddy, A. A. and Longton, J. J. (1969). *Inst. Brewing* **75**, 7.
Garcia Mendoza, C. G. and Villanueva, J. R. (1963). *Can. J. Microbiol.* **9**, 141.
Garcia Mendoza, C. G. and Villanueva, J. R. (1967). *Biochim. biophys. Acta* **135**, 189.
Garcia Mendoza, C. G., Garcia Lopez, M. D., Uruburu, F. and Villanueva, J. R. (1968). *J. Bact.* **95**, 2393.
Gascón, S. and Lampen, J. O. (1968). *J. biol. Chem.* **243**, 1567.
Gascón, S. and Ottolenghi, P. (1967). *C. r. Trav. Lab. Carlsberg* **36**, 85.
Lampen, O. J. (1966). Proc. 16th Symp. Soc. General Microbiol., p. 111.
Lampen, O. J. (1968). *Antonie van Leeuwenhoek* **34**, 1.
Lampen, O. J., Neumann, N. P., Gascón, S. and Montenecourt, B. S. (1967). *In* "Organizational Biosynthesis", (H. J. Vogel, J. O. Lampen and V. Bryson, eds.), p. 363, Academic Press, London and New York.
Longley, R. P., Rose, A. H. and Knights, B. A. (1968). *Biochem. J.* **108**, 401.
Matile, Ph. (1970). Proc. 2nd Symp. on Yeast Protoplasts, Brno 1968 (in press).
Matile, Ph., Moor, H. and Mühlethaler, K. (1967). *Arch. Mikrobiol.* **58**, 201.
Moor, H. and Mühlethaler, K. (1963). *J. Cell Biol.* **17**, 609.

Stahl, E. (1962). "Dünnschichtchromatographie", Springer-Verlag, Berlin, Göttingen and Heidelberg.
Stoffel, W., LeKim, D. and Sticht, G. (1968). *Hoppe-Seyler's Z. physiol. Chem.* **349**, 664.
Suomalainen, H., Nurminen, T. and Oura, E. (1967). *Acta chem. fenn.* B **40**, 323.
Wagner, H. and Zofcsik, W. (1966). *Biochem. Z.* **346**, 333.

The Resolution and Properties of Some Major Components of *Micrococcus lysodeikticus* Cell Membranes

E. MUÑOZ*, M. R. J. SALTON and D. J. ELLAR†

Department of Microbiology, New York University School of Medicine, New York, N.Y., U.S.A.

INTRODUCTION

The cell membranes of bacterial origin are now being studied in several laboratories (for a review, see Salton, 1967a). The cell membranes from gram-positive bacteria are easier to isolate than those of gram-negative bacteria. The most common method used for isolation of bacterial membranes involves the use of cell-wall degrading enzymes followed by the separation of the membranes from the cytoplasm upon differential centrifugation. Although criteria for establishing the homogeneity of cell membrane preparations have not been adequately investigated in the past, some progress is being made in defining the isolated membranes (Salton, 1967b). The chemical composition of the bacterial membranes is very similar to that of other membrane systems (Salton, 1967a), and a great deal is now known about the nature of the lipid components of these cell membranes (Kates, 1966). Although many biochemical activities have been found to be associated with the membrane fractions isolated from gram-positive bacteria (Salton, 1967a), very little is known about the protein constituents of these organelles and about the molecular anatomy of bacterial cell membranes.

In order to gain some insight into the molecular architecture and functioning of a bacterial cell membrane type, investigations in this laboratory have been directed towards the isolation and characterization of specific functional markers. The present report will review our studies on the resolution and properties of some components of the electron transport chain and an adenosine triphosphatase activity of *Micrococcus lysodeikticus* cell membranes.

---

* Present address: Instituto de Biologia Celular, C.S.I.C., Velazquez 144, Madrid-6, Spain.
† Present address: Biochemistry Department, University of Cambridge, England.

MATERIAL AND METHODS

Micrococcus lysodeikticus (NCTC 2665) was used in these studies. The conditions for growth and the procedure for membrane isolation were as described by Salton and Freer (1965) for directly lysed cells or as described by Muñoz *et al.* (1968a,b, 1969) when membranes were obtained by lysis of preformed, stabilized protoplasts. The membranes were then washed according to the "standard" procedure outlined by Salton (1967c) or washed four times with 0.03 M Tris, pH 7.5, for the membranes obtained from protoplasts (Muñoz *et al.,* 1968b, 1969). In both cases, membrane preparations were very similar and were called "standard" membranes.

The reagents, analytical procedures and enzyme assays have been previously described (Muñoz *et al.*, 1969; Ellar *et al.*, in preparation). Electron microscopy and glutaraldehyde fixation of the membranes were performed as indicated by Ellar *et al.* (in preparation).

RESULTS

The characteristics of our "standard" membrane preparations are summarized in Table 1.

Table 1. Characteristics of "standard" cell membranes isolated from *Micrococcus lysodeikticus*

	Per cent		References
Total membrane protein	100		
Total cellular protein	20-25		Salton and Freer, 1965; Muñoz *et al.*, 1968b
Protein	2.7*		
Lipid	2.0†		Salton, 1967c
Total enzymatic activities	Cytochromes	(> 94)	
	Succinic dehydrogenase	(> 94)	Ellar *et al.*, in preparation
	NADH$_2$-dehydrogenase	(> 80)	Dr. Nachbar, personal communication
	ATPase‡	(> 90)	Muñoz *et al.*, 1968b, 1969
Structure	Complex, membrane surface covered with granule-like structures		Ellar *et al.*, in preparation

* Protein determined by the Biuret method.
† Protein measured by the Lowry method.
‡ The membrane-bound ATPase cannot be totally measured. Calculated on the basis of maximal ATPase units after solubilization (Muñoz *et al.*, 1968b, 1969).

Deoxycholate extraction

By washing freshly isolated "standard" membranes with 0.05 M Tris-buffer, pH 7.5, containing 1% sodium deoxycholate (w/v), followed by centrifugation at 0-2°C for 50 min at 30,000 x g, Salton et al. (1968) have isolated a pellet of the characteristics listed in Table 2. As these results show, all the cytochromes and a great majority of the succinic dehydrogenase activity are localized in a defined structural entity representing only a small percentage of the "standard" membrane protein and, as a consequence, of the total cellular protein. This localization seems to be a good indication of the presence of these activities in a "base-piece" of the *Micrococcus lysodeikticus* cell membranes and a possible

Table 2. Properties of the insoluble residue after deoxycholate extraction of "standard" cell membranes from *Micrococcus lysodeikticus*

	Per cent	References
Total membrane protein	11-15.5	Salton et al., 1968
Total cellular protein	2-4	
Protein / Lipid	20-25*	Salton et al., 1968
Total enzymatic activities	Cytochromes (> 90)	Salton et al., 1968
	Succinic dehydrogenase (> 85)	Salton et al., 1968
	$NADH_2$-dehydrogenase (< 1)	
	ATPase (< 1)	
	Clean membranous sheets, collapsed appearance	Salton et al., 1968

* Protein estimated by the Lowry method.

role of these proteins in the structural integrity of these bacterial membranes.

Preliminary studies with the "DOC-soluble" fraction by gel electrophoresis have shown the predominant presence in this fraction of the previously identified ATPase band (Muñoz et al., 1968b, 1969). It has also been possible to measure in these "DOC-soluble" fractions high levels of $NADH_2$-dehydrogenase activity (Dr. M. S. Nachbar, personal communication).

ATPase "solubilization": properties of the "soluble" ATPase and of ATPase-depleted residues.

As indicated in Table 1, the "standard" membranes of *Micrococcus lysodeikticus* contained more than 90% of the "potential" ATPase present in this organism. We use the term "potential" ATPase because the membrane-bound protein does not show ATP hydrolysis unless trypsin activated (Muñoz et al., 1969). In our preliminary studies with the membrane-bound enzyme (Muñoz et al., 1969) there is no highly specific divalent cation dependence for the trypsin-activated

ATPase. The ATPase could be selectively released from its "standard" membrane complex by two well-defined steps: (i) prior cleaning of membranes by washing, and (ii) subjection of the membranes to a low ionic environment. Most of the results obtained in these studies have been reported in a previous communication (Muñoz et al., 1968b). When the ATPase is in a "soluble" state, it shows a more specific dependence for Ca^{2+} as an activator (Muñoz et al., 1969).

The "soluble" Ca^{2+}-dependent ATPase has been purified by gel filtration on Sephadex G-200. The progressive purification of this activity has been demonstrated by examining the various states of the preparation by gel electrophoresis, sedimentation velocity experiments, and by gel diffusion against membrane antiserum (Muñoz et al., 1969). The protein has been identified as a major component of *Micrococcus lysodeikticus* cell membranes (10% of the "standard" membrane protein) and ATPase-rich preparations showed in the electron microscope after negative staining, the presence of particles of ca. 100 Å diameter, possessing a central subunit surrounded by six additional subunits (Muñoz et al., 1968a). An indirect confirmation of the association between ATPase activity and this type of particle was obtained by electron microscopic observation of the ATPase-depleted membranes. In this case, smaller, less electron dense fragments, with almost complete absence of "structured" particles were observed (Ellar et al., in preparation).

It is interesting to note that the presence of 1mM EDTA in the washing fluids did not dissociate the ATPase protein from its membrane complex (Muñoz et al., 1968b), although it completely solubilized the $NADH_2$-dehydrogenase activity (Ellar et al., in preparation). The EDTA effect is not as yet totally understood and needs more careful study, but as shown in Table 3 it will result in a protein + lipid solubilization. Table 3 summarizes the properties of the residual pellets obtained after ATPase "solubilization" and the accompanying $NADH_2$-dehydrogenase release by action of dilute buffer with and without prior treatment of the membranes with EDTA. The cytochromes and succinic dehydrogenase activity always remained associated with the membranous structures obtained by these treatments.

$NADH_2$-dehydrogenase "solubilization"

Our studies with this activity are far less complete than those achieved with the ATPase activity. In general, we can say that the $NADH_2$-dehydrogenase is a membrane-bound enzyme whose membrane dissociation conditions are extremely dependent upon the isolation procedure. In contrast with ATPase, $NADH_2$-dehydrogenase association with the membrane seems to be directly dependent on divalent cations. The presence of EDTA in the washes resulted in a complete "solubilization" of this activity, but the same was not true for the ATPase protein. The action of dilute buffer did not dissociate a great amount of this dehydrogenase activity, unlike the ATPase activity that was specifically "solubilized" by this treatment. It is important to emphasize that the $NADH_2$-dehydrogenase dissociation pattern from the membranes is highly dependent on the conditions for membrane isolation.

Table 3. Properties of the ATPase-depleted residues after low ionic strength buffer (with and without prior EDTA) treatment

	A (− EDTA)	B (+ EDTA)
Per cent of total membrane protein	60-64	35-37
Per cent of total cellular protein	12-16	7-9
Protein / Lipid	1.5*	1.5*
Enzymatic activities	Cytochromes (+++) Succinic dehydrogenase (+++) $NADH_2$-dehydrogenase (+)	Cytochromes (+++) Succinic dehydrogenase (+++)
Structure	Smooth sheets, almost complete absence of "structured particles"	Smooth sheets, small fragments, absence of granule-like particles

* Protein estimated by the Lowry method.

Glutaraldehyde effect on enzyme solubilization and membrane structure
In view of the "solubilization" of membrane components, it was of interest to determine if the binding of these components could be modified in the membrane prior to isolation, to the extent that their release could be prevented, and if this would be accompanied by changes in the membrane appearance that could confirm the association existing between the "detachable" proteins and the particles seen in the "standard" membranes after negative staining.

If the membranes were obtained by the protoplasts state but treated with 0.5% glutaraldehyde by adding it in the bursting buffer (Ellar et al., in preparation), the washing with EDTA and dilute buffer resulted in a complete difference as compared with non-glutaraldehyde-treated membranes. In effect, with membranes obtained from glutaraldehyde-fixed protoplasts, only 10% of the $NADH_2$-dehydrogenase activity was released upon washing. Moreover, 90% of the activity remained in the residual membrane pellet. ATPase release was also inhibited by the glutaraldehyde treatment and a substantial amount of trypsin-activated ATPase was associated with the 0.5% glutaraldehyde-treated membrane, although quantitative assay was difficult owing to the complexity of trypsin activation (Muñoz *et al.*, 1969).

The fixed membrane after EDTA and dilute buffer action still showed in the electron microscope a structure close to the freshly isolated membranes, unlike the non-treated membranes (see above and Tables 1 and 3). The properties of the residues obtained after these various treatments are summarized in Table 4.

Table 4. Properties of the 0.5% glutaraldehyde-treated membranes after EDTA and low ionic strength buffer washing

Per cent of total membrane protein	50-60
Per cent of total cellular protein	11-13
Protein / Lipid	not done
Per cent of total enzymatic activities	Cytochromes ($>$ 94)
	Succinic dehydrogenase ($>$ 94)
	$NADH_2$-dehydrogenase ($>$ 85)
	ATPase (+++)*
Structure	Complex with associated granule-like particles (structure similar to "standard" membranes)

* Difficult to quantitate owing to the properties of the membrane-bound enzyme.

It is interesting to compare these results with those of column B of Table 3. The EDTA and dilute buffer-washed membranes contain, after glutaraldehyde fixation, 50% more protein than equally washed but untreated membranes. They also contain high levels of $NADH_2$-dehydrogenase and ATPase activities, and these changes in enzyme "solubilization" are paralleled by changes in the morphology, presenting a more complex structure. The structure of these residues with associated particles provides indirect support of the idea that ATPase, and probably other "detachable" proteins, are related to the organized particles seen in the electron micrographs of *Micrococcus lysodeikticus* cell membranes. The results obtained with glutaraldehyde fixation pose a new question about the structural and functional roles of the 40-50% of "standard" membrane protein "solubilized" by EDTA and dilute buffer action on the fixed membranes with no detectable changes on both aspects of the treated membranes.

DISCUSSION

The results reported here show that cytochromes and succinic dehydrogenase activity can be isolated by mild treatments or by detergent action in a defined structural entity of membranous character. If these catalytic properties are so intimately associated with an organized element of the cell membranes of *Micrococcus lysodeikticus*, there is a good indication that they play a role in the structural organization of the cellular membrane of this micro-organism. At the same time, the different treatments described by us have shown the existence of a group of "detachable" proteins (ATPase, $NADH_2$-dehydrogenase) probably associated with the granular sub-structures present in the "standard" membranes. Direct evidence in favor of this idea has been obtained by electron microscope observation of ATPase-rich fractions (Muñoz et al., 1968a), while indirect evidence has been presented by studying cell membranes of *Micrococcus lysodeikticus* treated with glutaraldehyde (Ellar et al., in preparation). These facts suggest a common type of structural organization for mitochondrial and some bacterial membranes. Our results are in contrast with those of Gelman and colleagues (Simakova et al., 1968) concluding that ATPase is localized in the stroma and not in the structured particles of *Micrococcus lysodeikticus*. Their conclusions were based upon unchanged ATPase activity and partial removal of the particles by proteolytic digestion of the isolated membranes. A possible explanation for this discrepancy could result from the sum of two facts: (i) the low ATPase activity of the isolated membranes, and (ii) the progressive unmasking of the ATPase by proteolysis that would compensate for the partial removal of the ATPase-containing particles. Further work needs to be done in this direction, but we hope that the understanding at a molecular level of a functional bacterial membrane has now been made clearer.

SUMMARY

The cell membranes of *Micrococcus lysodeikticus* obtained under defined conditions have been fractionated by using chemical agents. The fractionation of the membranes has been followed by biochemical and electron microscope techniques. The biochemical activities studied in our work were: cytochromes, succinic dehydrogenase, $NADH_2$-dehydrogenase and adenosine triphosphatase.

The results reported here lead us to conclude that cytochromes and succinic dehydrogenase remained associated with an organized structural entity, while $NADH_2$-dehydrogenase and adenosine triphosphatase could be dissociated from their membrane-complex. The "soluble" protein showing Ca^{2+}-dependent adenosine triphosphatase activity had been purified and some of its properties studied.

The evidence obtained in our studies seems to indicate an association between the ATPase protein and the particles seen by the electron microscopic observation of *Micrococcus lysodeikticus* cell membranes after negative staining.

ACKNOWLEDGMENTS

This work was supported by a National Science Foundation Grant (GB 7250). We are indebted to Dr. M. S. Nachbar for interesting information prior to publication and to Dr. J. H. Freer for his interest and his help during the electron microscope part of the work.

REFERENCES

Ellar, D. J., Muñoz, E. and Salton, M. R. J., in preparation.
Kates, M. (1966). *A. Rev. Microbiol.* **20**, 13.
Muñoz, E., Freer, J. H., Ellar, D. J. and Salton, M. R. J. (1968a). *Biochim. biophys. Acta* **150**, 531.
Muñoz, E., Nachbar, M. S., Schor, M. T. and Salton, M. R. J. (1968b). *Biochem. biophys. Res. Commun.* **32**, 539.
Muñoz, E., Salton, M. R. J., Ng, M. H. and Schor, M. T. *Eur. J. Biochem.* **7**, 490.
Muñoz, E., Salton, M. R. J., Ng, M. H. and Schor, M. T. (1969). *Eur. J. Biochem.* **7**, 490.
Salton, M. R. J. (1967a). *A. Rev. Microbiol.*, **21**, 417.
Salton, M. R. J. (1967b). *In* "The Specificity of Cell Surfaces", (B. D. Davis and L. Warren, eds.), p. 71, Prentice Hall, Englewood Cliffs, N.J.
Salton, M. R. J. (1967c). *Trans. N.Y. Acad. Sci.* Ser. II, **29**, 764.
Salton, M. R. J. and Freer, J. H. (1965). *Biochim. biophys. Acta* **107**, 531.
Salton, M. R. J., Freer, J. H. and Ellar, D. J. (1968). *Biochem. biophys. Res. Commun.* **33**, 909.
Simakova, I. M., Lukoyanova, M. A., Biryuzova, V. I. and Gelman, N. S. (1968). *Biokhimiya* **33**, 1047.

Biochemistry of the Bacterial Wall Peptidoglycan in Relation to the Membrane

J. M. GHUYSEN and M. LEYH-BOUILLE

Service de Bactériologie, Université de Liège, Belgium

The wall peptidoglycan and the inner plasma membrane are distinct but functionally interdependent organelles. Essentially, the peptidoglycan is a supporting structure which protects the membrane against deleterious influences. In particular, it allows the bacteria to live under environmental conditions which are usually hypotonic. Conversely, the membrane is actively involved in the peptidoglycan synthesis and, probably, in the functioning of several autolytic enzymes (that are possible factors of wall synthesis and regulation), as well as in the enzymatic inactivation of several antibiotics such as benzylpenicillin (which is known to inhibit the peptidoglycan synthesis).

During bacterial growth and division, the peptidoglycan layer undergoes many reactions, i.e. the creation of receptor sites by hydrolysis and the insertion of newly synthesized building blocks. In spite of the fact that the peptidoglycan is one of the most dynamic structures of the whole cell, its mechanical strength is never impaired, at least under well-balanced growth conditions. The biochemical expression of this remarkable property resides in the structure of the peptidoglycan, i.e. a network of glycan strands interlinked through peptide chains.

Essentially, the glycan chains are composed of alternating β-1,4-linked N-acetylglucosamine and N-acetylmuramic acid pyranoside residues, i.e. a chitin-like structure, except that every other sugar is substituted by a 3-O-D-lactyl group and the average chain length is small (20 to 140 hexosamine residues, according to the bacterial species). The peptide chains are composed of tetrapeptide subunits which substitute through their N-termini the glycan D-lactic acid groups and of peptide bridges which cross-link the tetrapeptide subunits of adjacent glycan chains. The peptide moiety is a small size component with an average size of 1.5 to 10 cross-linked peptide subunits. Many terminal groups thus occur in both the glycan and the peptide parts of the peptidoglycan net, as it is isolated, which probably reflect the dynamics of the bacterial growth.

With the possible exceptions that the residue at the amino terminus is not always L-alanine but sometimes L-serine or glycine, and that D-glutamic acid may be replaced by a derivative of it, such as 3-hydroxyglutamic acid, the tetrapeptide subunits have the general sequence:

(L)
L-Ala-γ-D-Glu-CH-D-Ala (Fig. 1).
|
X

Amide ammonia, glycine and D-serine may substitute the α-COOH group of D-glutamic acid. Amide ammonia may also substitute the free COOH group of diaminopimelic acid.

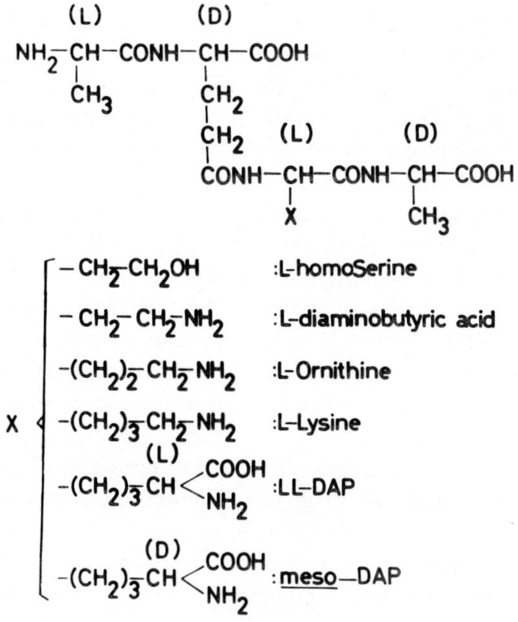

```
         (L)        (D)
    NH₂-CH-CONH-CH-COOH
        |           |
        CH₃         CH₂
                    |
                    CH₂   (L)        (D)
                    |
                    CONH-CH-CONH-CH-COOH
                         |           |
                         X           CH₃
```

$$X \begin{cases} -CH_2-CH_2OH & :\text{L-homoSerine} \\ -CH_2-CH_2-NH_2 & :\text{L-diaminobutyric acid} \\ -(CH_2)_2-CH_2-NH_2 & :\text{L-Ornithine} \\ -(CH_2)_3-CH_2-NH_2 & :\text{L-Lysine} \\ -(CH_2)_3-\overset{(L)}{CH}\underset{NH_2}{\overset{COOH}{<}} & :\underline{\text{LL-DAP}} \\ -(CH_2)_3-\overset{(D)}{CH}\underset{NH_2}{\overset{COOH}{<}} & :\underline{\text{meso}}-\text{DAP} \end{cases}$$

Figure 1. General structure of the tetrapeptide subunits. The L configuration of homoserine in *C. poinsettiae* has been proved by Dr. H. R. Perkins (personal communication).

Four types of cross-linking between the tetrapeptide subunits are known (Ghuysen, 1969). They represent all the possible basal variations unless unexpected mechanisms in the peptidoglycan synthesis were still to be discovered. The cross-linking always involves the C-terminal D-alanine of one peptide subunit and either the free amino group of the diamino acid (types I, II, III) or the α-carboxyl group of D-glutamic acid (type IV) of another peptide subunit. A type IV variant, which has not yet been encountered, might involve the carboxyl group of diamino-pimelic acid instead of the α-carboxyl group of D-glutamic acid.

Type I (*Escherichia coli*; probably all the Gram negative bacteria) (Fig. 2) is characterized by a direct bond involving the amino group located on the D-carbon of *meso*-diaminopimelic acid, that is to say a D-alanyl-(D)-*meso*-diaminopimelic acid linkage (van Heijenoort *et al.*, 1969). Type II (some *Micrococcus* sp., *Staphylococcus*, *Gaffkya*, *Streptococcus*, *Lactobacillus*, *Leuconostoc*, *Streptomyces* sp., *Welchia perfringens*) (Fig. 3) consists in a single

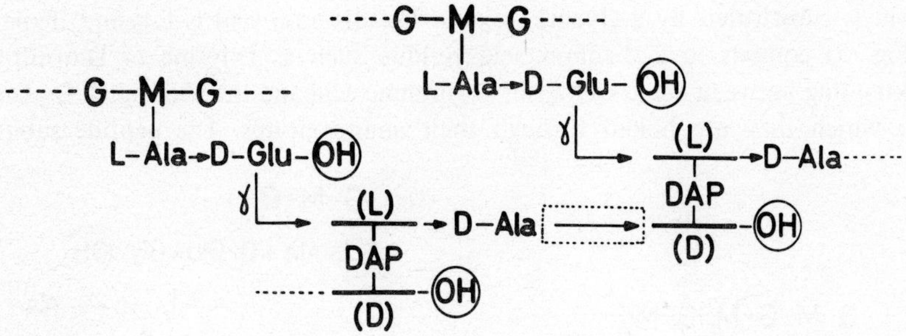

Figure 2. Peptidoglycan of type I. G = *N*-acetylglucosamine; M = *N*-acetylmuramic acid.

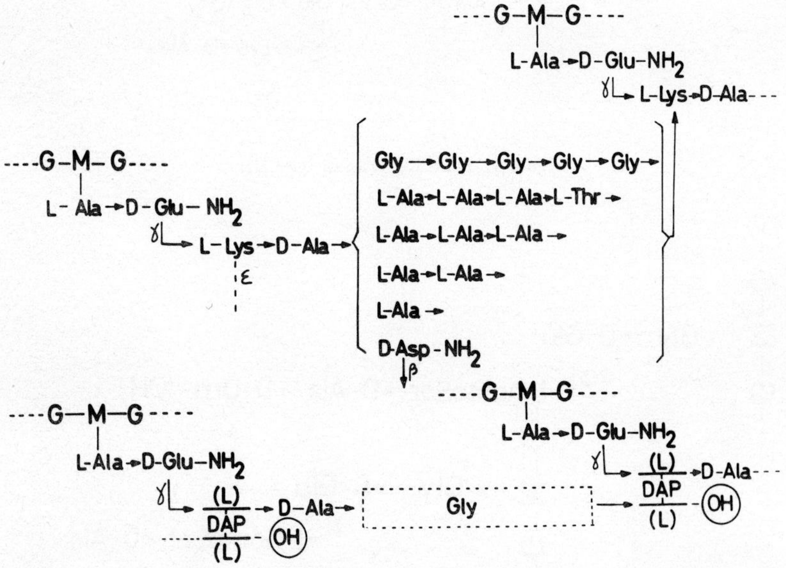

Figure 3. Peptidoglycan of type II. Upper part: see *Bact. Rev.* 1968, 32, 425. A pentaglycine bridge is typical of *Staphylococcus aureus*. Lower part: peptidoglycan of a *Streptomyces* sp. (R. Bonaly and J. M. Ghuysen, unpublished results) and of *Welchii perfringens* (Leyh-Bouille *et al.*, 1970a).

additional amino acid (glycine, L-amino acid, D-isoasparagine) or in a peptide chain (of glycine and/or L-amino acid residues) containing up to five residues, extending between the C-terminal D-alanine and the free amino group of the diamino acid (i.e. L-ornithine, L-lysine or LL-diaminopimelic acid). The α-COOH of the D-glutamic acid is often, if not always, amidated. Type III (*Micrococcus lysodeikticus* and related *Micrococci*) (Fig. 4) is similar to type II except that the peptide cross-linking consists in one or several peptides (interlinked through D-alanyl-L-alanine linkages) each having the same amino acid sequence as the peptide subunit (Campbell *et al.*, 1969), and that the α-COOH of D-glutamic acid is substituted by a glycine residue. The diamino acid is L-lysine. Type IV (Fig. 5) consists in a diamino acid residue such as D-lysine or D-ornithine, extending between the α-COOH of D-glutamic acid and the C-terminal D-alanine to which they are linked through their amino groups. The peptide subunits

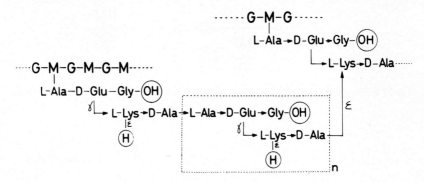

Figure 4. Peptidoglycan of type III.

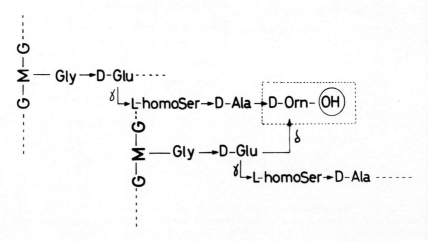

Figure 5. Peptidoglycan of type IV (from Perkins, H. R., 1965).

present rather unusual sequences: N^α-(L-Ser-γ-D-Glu)-L-Orn-D-Ala (*Butyribacterium rettgeri*) (Guinand et al., 1969) and Gly-γ-D-Glu-homoSer-D-Ala (plant pathogenic *Corynebacteria*).

Electron microscopy studies strongly suggest that the peptidoglycan sheet in the Gram negative bacteria is a 20 Å-thick network (peptidoglycans of type I). The peptidoglycan layer in the Gram positive *Bacillaceae* is about 100 Å-thick although it is composed of the same *meso*-diaminopimelic acid peptide subunits as those of *E. coli* (van Heijenoort et al., 1969). In one case, that of *B. megaterium* KM, not only *meso* but also DD-diaminopimelic acid residues are present. One may hypothesize that the DD-residues serve to interconnect several superimposed type I, *E. coli*-like, peptidoglycan monolayers. Similarly, in *Micrococcus varians*, the *meso*-diaminopimelic acid-containing peptide subunits appear to be cross-linked through polyglutamic acid bridges (with their amino termini linked to COOH groups of D-alanine) so that the *M. varians* peptidoglycan would actually belong to type II (M. Leyh-Bouille and J. M. Ghuysen, unpublished results).

Several particulate enzymes and a C_{55} polyisoprenoid alcohol monophosphate, which are believed to be parts of the membrane or, at least, to be located on it, are involved in a complex, multiple stage reaction (Fig. 6) which, according to Strominger and his colleagues (Strominger, 1969), deals with: (i) the assembly of the two precursors uridine-5'-pyrophosphoryl-*N*-acetylglucosamine (UDP-GlcNAc) and uridine-5'-pyrophosphoryl-*N*-acetylmuramylpentapeptide:

$$\text{(UDP-MurNAc-R}_1\text{-}\gamma\text{-D-Glu-}\underset{\underset{\text{X}}{|}}{\overset{\text{(L)}}{\text{CH}}}\text{-D-Ala-D-Ala)}$$

into β-1,4-GlcNAc-MurNAc-pentapeptide units. In this process, first MurNAc (pentapeptide)-monophosphate is transferred from the nucleotide precursor to the P-C_{55} lipid resulting in the liberation of UMP and the attachment of MurNAc-pentapeptide to the lipid by means of a pyrophosphate bridge. Next, GlcNAc is transglycosylated from UDP-GlcNAc with liberation of UDP and formation of disaccharide (pentapeptide)-P-P-C_{55} lipid. (ii) Possible modifications of the pentapeptide units, such as the amidation of carboxyl groups, the substitution of the α-COOH of D-glutamic acid by an additional amino acid, or the incorporation of those amino acid residues which in the completed peptidoglycan will play the role of "specialized" peptide bridges (types II, III and IV). Usually, the type II bridges incorporation is aminoacyl-tRNA-dependent, though ribosomes seem not to be involved in the process. The mechanisms involved in the incorporation of type III and IV bridges are

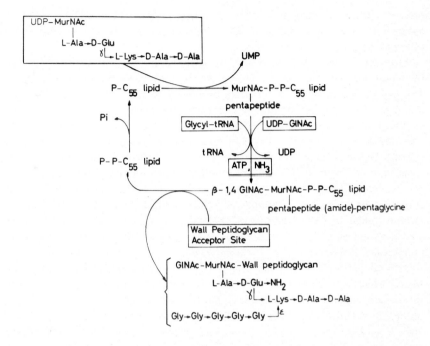

Figure 6. The lipd cycle in *Staphylococcus aureus* with formation of a nascent uncross-linked peptidoglycan (from Strominger, 1969). In *Escherichia coli*, the same cycle occurs except that D-glutamic acid is not amidated and no additional amino acids are incorporated in the pentapeptide moiety.

unknown. (iii) The transfer to the growing wall peptidoglycan of the newly synthesized disaccharide peptide units and their insertion into wall receptor sites. During this transfer, C_{55} lipid pyrophosphate is liberated and dephosphorylated so that the P-C_{55} carrier can begin a new cycle. The wall receptor sites are probably non-reducing N-acetylglucosamine termini in the glycan chains. Endo-N-acetylmuramidase autolysines may be involved in the creation of such receptors. (iv) The closure of the bridges between the peptide subunits so that the nascent peptidoglycan is being transformed into a rigid two- or three-dimensional network (Fig. 7). This last reaction would be a transpeptidation in which the bond energy of the terminal D-alanyl-D-alanine of one peptide subunit is utilized to transfer the carboxyl group of the penultimate D-alanine residue to the amino acceptor of a second peptide subunit. Concomitantly, the terminal D-alanine is released. The end product is always an insoluble network of glycan chains substituted by cross-linked tetrapeptides and the C-terminal D-alanine residue of these tetrapeptides are always involved in the cross-linking.

None of the membrane-bound enzymes involved in the lipid cycle and bridge

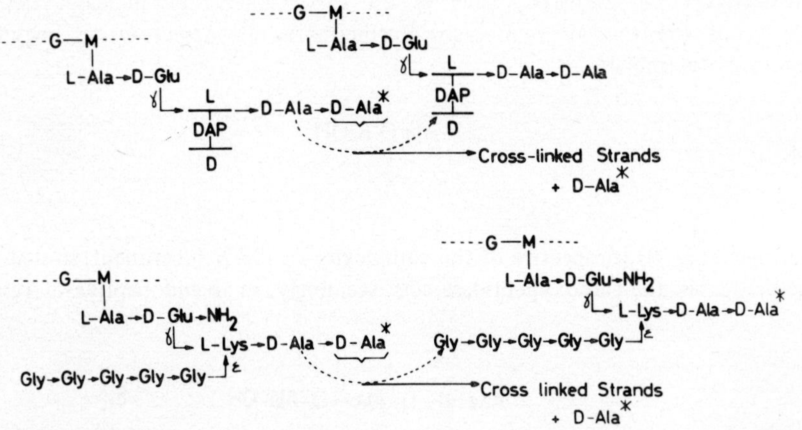

BIOCHEMISTRY OF THE BACTERIAL WALL PEPTIDOGLYCAN 65

Figure 7. The bridge closure reaction in *Escherichia coli* (upper part, compare with Fig. 2) and in *Staphylococcus aureus* (lower part, compare with Fig. 3) (Strominger, 1969). A direct demonstration of the transpeptidation reaction has only been provided in the case of *Escherichia coli* (Izaki et al., 1968).

closure reactions of the peptidoglycan synthesis pathway has been isolated and purified to the stage of enzymatic homogeneity. Recently, however, a D-alanine carboxypeptidase which releases the terminal D-alanine residue from L-Ala-γ-D-Glu-(L)-*meso*-DAP-(L)-D-Ala-D-Ala was detected in a particulate fraction of *E. coli* (Izaki et al., 1968). It was partially purified in the form of a soluble enzyme (Izaki and Strominger, 1968) and was said to be involved in the regulation of the size of the peptide moiety of the *E. coli* peptidoglycan which is known to be very poorly cross-linked. Indeed, the removal at some stage of the peptidoglycan synthesis of the terminal D-alanine residues from some pentapeptides, consequently decreases the number of peptides susceptible to undergo the bridge closure reaction. A similar soluble D-alanine carboxypeptidase was isolated from a *Streptomyces* culture filtrate. In contrast to *E. coli*, the *Streptomyces* peptidoglycan belongs to type II, a single glycine residue being involved in the cross-linking between L-Ala-γ-D-Glu (amide)-(L_1)-LL'-DAP-(L_1)-D-Ala tetrapeptides (Fig. 3). Preliminary results* reveal that the release by the *Streptomyces* carboxypeptidase of the terminal D-alanine residue from pentapeptides L-Ala-γ-D-Glu-R_3-D-Ala-D-Ala is essentially controlled by the structure of the residue or chemical group at the R_3 position (Fig. 8). N^α-(L-Ala-γ-D-Glu)-L-Lys-D-Ala-D-Ala is hydrolyzed at a very slow rate. The replacement of L-lysine by N^ϵ-(D-isoasparaginyl)-L-lysine, homoserine, *meso*-diaminopimelic acid and N^ϵ-

---

* This research was initiated in collaboration with the Institut de Biochimie, Laboratoire des Peptides, Orsay, France (Dr. E. Bricas) (van Heijenoort et al., 1969). A full report will be presented elsewhere (Ghuysen et al., 1970; Leyh-Bouille et al., 1970b).

(pentaglycyl)-L-lysine is paralleled by a 6, 15, 20 and 80-fold increase, respectively, of the rate of hydrolysis. Furthermore, the *Streptomyces* enzyme hydrolyses terminal:

$$\text{D-Ala-D-R-OH} \atop | \atop \text{X}$$

linkages (Fig. 9) irrespective of the complexity of the X substituent, so that, in several cases, the carboxypeptidase acts, seemingly, as an endopeptidase. Again,

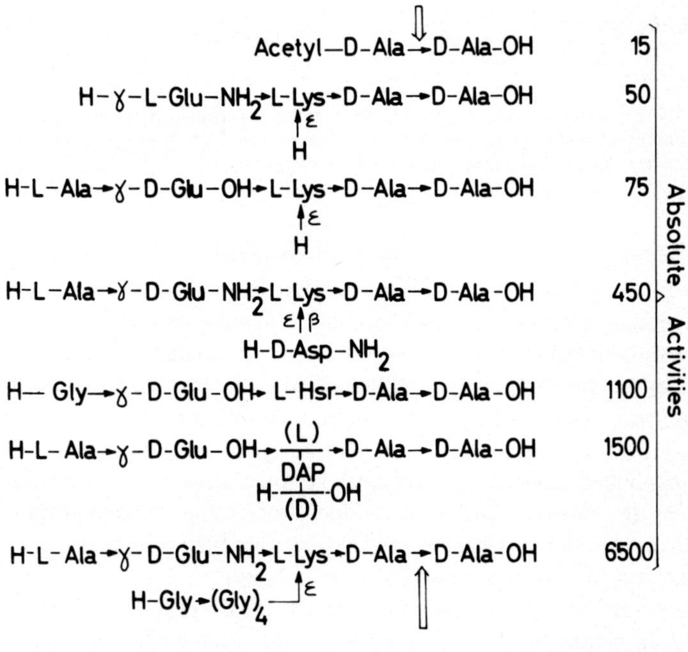

Acetyl—D-Ala→D-Ala-OH	15
H-ɣ-L-Glu-NH₂→L-Lys→D-Ala→D-Ala-OH ↑ε H	50
H-L-Ala→ɣ-D-Glu-OH→L-Lys→D-Ala→D-Ala-OH ↑ε H	75
H-L-Ala→ɣ-D-Glu-NH₂→L-Lys→D-Ala→D-Ala-OH ε↑β H-D-Asp-NH₂	450
H—Gly→ɣ-D-Glu-OH→L-Hsr→D-Ala→D-Ala-OH	1100
H-L-Ala→ɣ-D-Glu-OH→(L)/↓→D-Ala→D-Ala-OH DAP H—↓—OH (D)	1500
H-L-Ala→ɣ-D-Glu-NH₂→L-Lys→D-Ala→D-Ala-OH ↑ε H-Gly→(Gly)₄	6500

Absolute Activities

Figure 8. Substrate specificity of *Streptomyces* D-alanine carboxypeptidase. Action on D-alanyl-D-alanine linkages. Results (absolute activity) are expressed in nanoequivalents of hydrolyzed linkages/h/mg of protein. A typical assay mixture contains 15 nanoequiv. of D-Ala-D-Ala linkage, 1 to 10 μg of protein, in a total volume of 30 μl of 0.01 M Veronal buffer, pH 9. N^α-(UDP-MurNAc-L-Ala-γ-D-Glu)-L-Lys-D-Ala-D-Ala and UDP-MurNAc-Gly-γ-D-Glu-homoSer-D-Ala-D-Ala were gifts from Dr. H. R. Perkins. UDP-MurNAc-L-Ala-γ-D-Glu-(L)-*meso*-DAP-(L)-D-Ala-D-Ala was a gift from Dr. A. J. Garrett. These nucleotides were used to prepare the corresponding pentapeptides. Nucleotides and free pentapeptides have very similar enzymatic sensitivities. Origin of the peptides, from top to bottom: 1, synthetic; 2, synthetic; 3, *S. aureus*, nucleotide; 4, *L. acidophilus*, walls; 5, *C. poinsettiae*, nucleotide; 6, *B. subtilis* W23, nucleotide; 7, *S. aureus*, wall from cells treated with sublethal doses of penicillin.

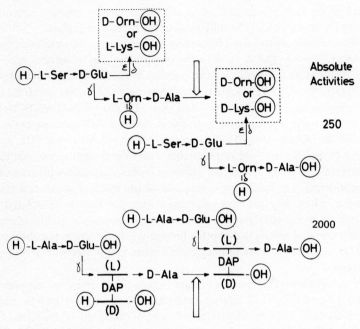

Figure 9. Substrate specificity of *Streptomyces* D-alanine carboxypeptidase. Action on D-Ala → D-R-OH linkages. Results and assays, see Fig. 8. Upper part: dimer from *B. rettgeri*; lower part: dimer from *E. coli*.

the influence of the R_3 residue, i.e. the residue immediately preceding the terminal D-Ala-D-R-OH sequence, seems to be of prime importance since the replacement of L-ornithine by *meso*-diaminopimelic acid results in a 80-fold increase of the rate of hydrolysis of the:

$$\text{D-Ala-D-R-OH linkages.} \atop \text{X}$$

All the foregoing examples are consistent with the hypothesis that the presence at the R_3 position of an ω-amino group (N^ϵ-L-lysine or N^δ-L-ornithine) which is not a possible acceptor for a bridge closure transpeptidation reaction, makes the D-alanyl-D-alanine and more generally the D-Ala-D-R-OH linkages fairly resistant to the carboxypeptidase. Conversely, the introduction at the R_3 position of homoserine (i.e. a neutral amino acid not involved in the transpeptidation; in *C. poinsettiae* the peptide cross-linking involves the α-amino group of a D-ornithine residue which substitutes the α-COOH of D-glutamic acid, see Fig. 5) or of an α-amino group that can be utilized in the bridge closure reaction, considerably enhances the sensitivity of the substrates to the carboxypeptidase.

To all appearances and from the scanty information so far available (Izaki and Strominger, 1968), the *E. coli* carboxypeptidase has substrate specificities similar to those of the *Streptomyces* enzyme. Particularly, UDP-MurNAc-L-Ala-γ-D-Glu-(L)-*meso*-DAP-(L)-D-Ala-D-Ala is again a much better substrate for the *E. coli* enzyme than N^α-(UDP-MurNAc-L-Ala-γ-D-Glu)-L-Lys-D-Ala-D-Ala. Moreover, the Michaelis constants of both the *E. coli* and the *Streptomyces* enzymes for UDP-MurNAc-L-Ala-γ-D-Glu-(L)-*meso*-DAP-(L)-D-Ala-D-Ala are virtually identical (6 and 8 x 10^{-4} M respectively). From the foregoing, one can visualize both enzymes as being able to act either as carboxypeptidases (or seemingly, as endopeptidases), i.e. catalyzing a reaction which involves water and results in hydrolysis, when they are in a solubilized form (as they were isolated), or as transpeptidases, i.e. catalyzing a reaction which involves a specific acceptor in place of water and results in α-peptide linkage synthesis, when, possibly, they are bound *in situ* to some lipoprotein constituents at the exterior of the plasma membrane.

The *Streptomyces* carboxypeptidase was isolated from an organism very resistant to penicillins and related antibiotics. *Streptomyces* growth inhibition required at least 500 μg/ml of Cephalothin, 1000 μg/ml of Penicillin G, and 10,000 μg/ml of Oxacillin. *In vitro* assays under the conditions described in Fig. 8 and using N^α-[GlcNAc-MurNAc-L-Ala-γ-D-Glu (amide)], N^ϵ-(Gly)$_5$-L-Lys-D-Ala-D-Ala as substrate, showed that a 50% inhibition of the carboxypeptidase activity was performed by cephalothin and penicillin G at molar ratios, antibiotic to substrate, of 15 and 70 to 1, respectively. Oxacillin at a molar ratio of 70 to 1, did not exhibit any inhibitory effect at all. Since the selected antibiotics are known to inhibit the transpeptidation reaction, the above parallelism between the *in vivo* and *in vitro* inhibition tests strengthens the hypothesis that the carboxypeptidase is indeed the transpeptidase which has undergone solubilization. All in all, however, much remains to be done to definitely ascertain our proposal.

In contrast to the *Streptomyces* enzyme, the *E. coli* carboxypeptidase is exceedingly sensitive to penicillins. Though they exhibit similar, perhaps identical, catalytic activities, the *Streptomyces* and the *E. coli* carboxypeptidases present quite different affinities for penicillins and related antibiotics. The fact that the *Streptomyces* "DD carboxypeptidase-transpeptidase" system is not inhibited by penicillins seems to be at variance with the idea (Tipper and Strominger, 1965) that an analogy between penicillins and the conformation of acyl-D-alanyl-D-alanine would be the molecular basis of the antibacterial action of these antibiotics. It shows, at least, that such an analogy is not universal among bacteria. (For a full discussion; see Leyh-Bouille *et al.*, 1970b).

SUMMARY

The peptidoglycan physically protects the membrane against deleterious influences. The membrane is involved in peptidoglycan synthesis and in other wall regulation mechanisms. The possible physiological function of a soluble carboxypeptidase active on acyl-D-alanyl-D-R-OH substrates and its relation to the membrane-bound transpeptidase involved in peptidoglycan synthesis are discussed.

ACKNOWLEDGMENTS

This investigation was supported by grant ER-E4-10-2 from the U.S. Department of Agriculture under Public Law 480 and by the Fonds de la Recherche Fondamentale Collective, Brussels, Belgium Contrats no. 515 and 1000.

REFERENCES

Campbell, J. N., Leyh-Bouille, M. and Ghuysen, J. M. (1969). *Biochemistry* 8, 193.
Ghuysen, J. M. (1968). *Bact. Rev.* 32, 425.
Ghuysen, J. M., Leyh-Bouille, M., Bonaly, R., Nieto, M., Perkins, H. R., Schleifer, K. H. and Kandler, O. (1970). *Biochemistry*, in press.
Guinand, M., Ghuysen, J. M., Schleifer, K. H. and Kandler, O. (1969). *Biochemistry* 8, 207.
Izaki, K. and Strominger, J. L. (1968). *J. biol. Chem.* 243, 3193.
Izaki, K., Matsuhashi, M. and Strominger, J. L. (1968). *J. biol. Chem.* 243, 3180.
Leyh-Bouille, M., Bonaly, R., Ghuysen, J. M., Tonelli, R. and Tipper, D. J. (1970a). *Biochemistry*, in press.
Leyh-Bouille, M., Ghuysen, J. M., Bonaly, R., Nieto, M., Perkins, H. R., Schleifer, K. H. and Kandler, O. (1970b). *Biochemistry*, in press.
Perkins, H. R. (1969). *Biochem. J.* 111, 195.
Strominger, J. L. (1969). *In* "Inhibitors, Tools in Cell Research" (Th. Bircher and H. Sies, eds.), p. 187. Springer-Verlag, Berlin.
Tipper, D. J. and Strominger, J. L. (1965). *Proc. natn. Acad. Sci. U.S.A.* 54, 1133.
van Heijenoort, J., Elbaz, L., Dezélée, Ph., Petit, J. F., Briscas, E. and Ghuysen, J. M. (1969). *Biochemistry* 8, 200.

The Mode of Action of Membrane-Active Antibacterials

W. A. HAMILTON

Department of Biochemistry, University of Aberdeen, Aberdeen, Scotland

We have been studying the mode of action of two synthetic antibacterials, TCS and TCC (Fig. 1). TCS is taken up by the cell membrane, and has been demonstrated to be an uncoupler of oxidative phosphorylation in bacterial and mitochondrial systems (Hamilton, 1968). Preliminary data show TCC to have similar properties. Accordingly, we have compared these two compounds with other antibacterials known to affect the membrane, e.g. CTAB (Salton, 1951), or to act as uncouplers, e.g. the polypeptide antibiotics gramicidin and valinomycin (Hotchkiss, 1944; Höfer and Pressman, 1966).

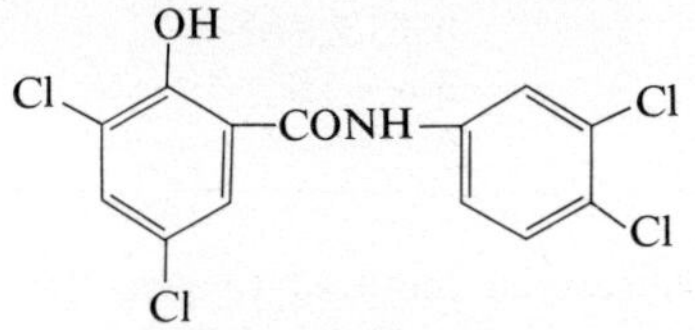

Tetrachlorosalicylanilide (TCS)

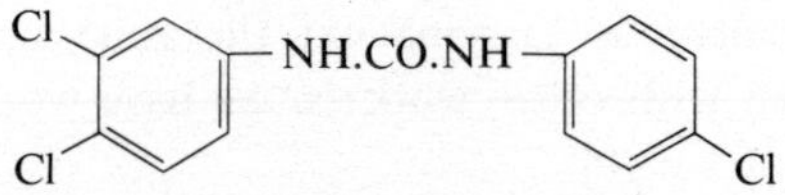

Trichlorocarbanilide (TCC)

Figure 1

These compounds are bacteriostatic to Gram positive organisms at low concentration, and very much less active against Gram negative species (Table 1). *Staphylococcus aureus* has been taken as our test organism.

Abbreviations: TCS, tetrachlorosalicylanilide; TCC, trichlorocarbanilide; CTAB, cetyl trimethylammonium bromide; GRAM, gramicidin; VAL, valinomycin; MIC, minimum inhibitory concentration.

Table 1. Minimum inhibitory concentration (µg/ml) with 10^7 cells/ml

	TCS	TCC	CTAB	GRAM
Staph. aureus	0.15	0.8	0.39	0.19
B. megaterium	0.20	0.20	0.39	0.19
M. lysodeikticus	0.20	0.12	0.19	0.09
E. coli	30	>100	12.5	>100
Ps. aeruginosa	>100	>100	>100	>100

We have previously stressed the importance of carrying out biochemical and other studies at the same relative concentrations of cells and antibacterial that cause the inhibition of growth. Using radioactively-labelled compounds, we can measure the molecules/bacterium (M/B) taken up at the MIC with an inoculum of 10^7 cells/ml (Table 2). In the last column we see the suspension or total

Table 2. Uptake of antibacterials by Staph. aureus

	MIC (C_T with 10^7 cells/ml) (µg/ml)	M/B	C_T (with 6×10^9 cells/ml to give same M/B) (µg/ml)
TCS	0.15	0.75×10^5	0.5
TCC	0.08	3.6×10^5	1.32
CTAB	0.39	6.3×10^6	24

concentration (C_T) of antibacterial required to give the same cellular concentration with a suspension containing 6×10^9 cells/ml, or its equivalent for *Staph. aureus* of 1 mg dry wt/ml. Unfortunately, the absence of labelled gramicidin and valinomycin has prevented us carrying out the same analysis for these compounds and, in parallel with other workers, we have worked at the arbitrarily chosen concentration of 1 µg/ml.

Although it was not possible in the present analysis of protoplast and spheroplast osmotic stability to use *Staph. aureus*, the same equivalent bacteriostatic conditions were applied, i.e. cell suspensions were prepared to a cell density of 1 mg/ml, protoplasts or spheroplasts were formed, and the effects of TCS, TCC, CTAB, gramicidin and valinomycin were studied at 0.5, 1.32, 24, 1 and 1 µg/ml respectively.

Protoplasts of *Bacillus megaterium* KM were formed by lysozyme treatment in 0.4 M sucrose, then centrifuged and resuspended in various hypertonic media. There was immediate lysis in 0.4 M glycerol or 0.5 M NH_4 acetate, but the protoplasts were stable in 0.5 M solutions of NH_4NO_3, NH_4Cl, KNO_3, KCl,

NaNO$_3$ and NaCl. From this it is taken that the protoplast membrane is permeable to glycerol and to NH$_4^+$ and acetate$^-$ ions (or NH$_3$ and acetic acid, (Chappell and Haarhoff, 1966)), but impermeable to K$^+$, Na$^+$, NO$_3^-$ and Cl$^-$ ions. That the protoplasts are stable in NH$_4$NO$_3$ and NH$_4$Cl is considered to be due to entry of NH$_4^+$ without concomitant movement of NO$_3^-$ or Cl$^-$, which establishes an electrical gradient inhibiting the further entry of NH$_4^+$.

The addition of CTAB to protoplasts stabilized in any of these media results in rapid lysis, i.e. CTAB destroys the membrane's semi-permeable character in a non-specific manner, independent of the medium composition.

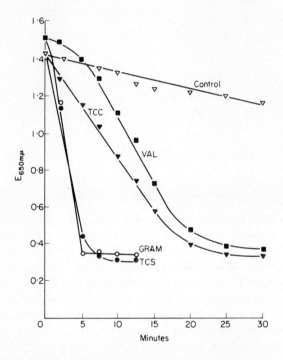

Figure 2. KM protoplasts in 0.5 M NH$_4$NO$_3$.

The effect of adding the other antibacterials to protoplasts stabilized in NH$_4$NO$_3$ is shown in Fig. 2. All four compounds, and especially TCS and gramicidin, increase the permeability of NO$_3^-$ with resultant lysis. However, in KNO$_3$ medium (Fig. 3) only those compounds already known to increase the K$^+$ permeability, i.e. gramicidin and valinomycin (Chappell and Crofts, 1965; Moore and Pressman, 1964; Harold and Baarda, 1967), cause protoplast lysis. In NaNO$_3$, lysis was only obtained with gramicidin, thus confirming the K$^+$ specificity of valinomycin.

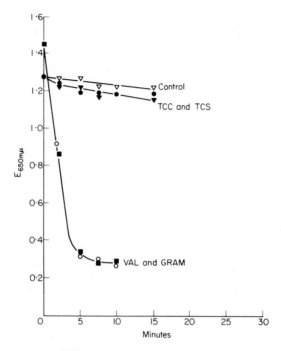

Figure 3. KM protoplasts in 0.5 M KNO_3.

There is a more complex response with protoplasts stabilized in NH_4Cl (Fig. 4). No effect is found with TCS, a slow rate of lysis is evident with valinomycin and gramicidin, a more rapid lysis with TCC, and the fastest rate is with TCC plus gramicidin. Our interpretation is that in addition to the natural permeability to NH_4^+ ions, TCC renders the membrane permeable to Cl^-. It appears, however, that in the untreated membrane the permeability to NH_4^+ is not maximal, nor the impermeability to Cl^- absolute, and that gramicidin and valinomycin can increase the NH_4^+ permeability. Hence the slow lysis with these antibiotics singly, and the increase in TCC-lysis with the addition of gramicidin.

These conclusions are reinforced by consideration of the effects in KCl (Fig. 5). As was shown with KNO_3, TCC does not affect K^+ permeability, and here again lysis is only found with gramicidin and valinomycin, either on their own (slow) or with TCC (fast). In NaCl (Fig. 6), lysis is only evident with gramicidin and gramicidin plus TCC.

In this study no membrane permeability, natural or induced, was found for Ca^{2+}, Mg^{2+}, Mn^{2+}, SO_4^{2-} or $H_2PO_4^-$. Also, no effect was noted in any medium with dinitrophenol at 1 µg/ml, 24 µg/ml or 5 mM. We can therefore describe the mode of the bacteriostatic action of these compounds in the following terms: CTAB destroys the membrane's semi-permeable character in a

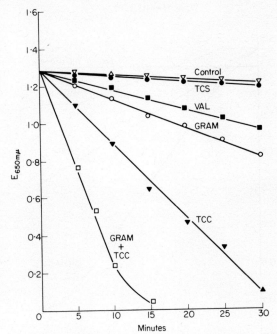

Figure 4. KM protoplasts in 0.5 M NH_4Cl.

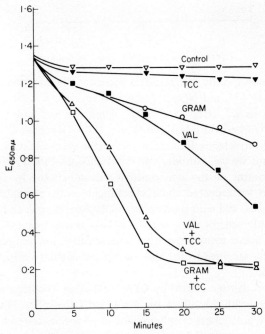

Figure 5. KM protoplasts in 0.5 M KCl.

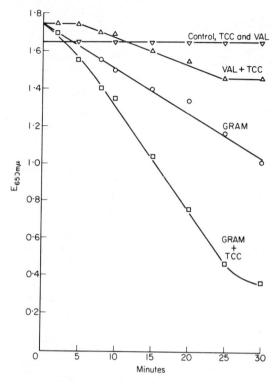

Figure 6. KM protoplasts in 0.5 M NaCl.

generalized and non-specific manner; TCS increases the membrane permeability to NO_3^-; TCC to Cl^- and NO_3^-; gramicidin to Na^+, K^+, NH_4^+ and NO_3^-; and valinomycin to K^+, NH_4^+ and NO_3^-. Dinitrophenol and TCS have been shown by other workers (Mitchell and Moyle, 1967; Hopfer et al., 1968; Harold and Baarda, 1968) to selectively increase membrane permeability to H^+.

From this point we can consider two developments. Firstly, what can we say about the mechanism of the non-sensitivity demonstrated by Gram negative organisms? When we repeated these experiments with protoplasts of *Micrococcus lysodeikticus*, and with spheroplasts of *Escherichia coli* and *Pseudomonas aeruginosa*, we obtained the same pattern of lysis as was demonstrated with KM protoplasts. The naked membranes from both sensitive and non-sensitive species are clearly equally sensitive to the specific ion permeability effects found with these antibacterials.

The generalized damage caused by CTAB allows us to compare its effect on protoplasts (lysis) with its effect on whole cells (leakage). Whereas lysis is obtained with the protoplasts or spheroplasts of all species tested in all media, at the same relative concentrations of antibacterial and cells, leakage from whole cells could only be demonstrated conclusively in those species sensitive to the bacteriostatic action of low concentrations of CTAB. This is demonstrated in Table 3 where leakage is measured by E_{260} mμ of the cell-free supernatant after treatment with CTAB for 1 h at 0°C. We can compare with the pool material extracted by treatment of the cells for 10 min in a boiling water bath. Even when autolysis takes place by treatment at 37°C, the leakage from *E. coli* still only represents 38% of the total pool material, and that from *Ps. aeruginosa* only 22%.

Table 3. Leakage from whole cells with CTAB (24 μg/ml)
(ΔE_{260} mμ of cell-free supernatant against untreated control)

	Staph. aureus	B. megaterium	M. lysodeikticus	E. coli	Ps. aeruginosa
100°C for 10 min	0.41	0.79	0.65	1.90	1.47
CTAB at 0°C for 1 h	0.21	0.17	0.13	0.04	0.00
CTAB at 37°C for 1 h	0.50	0.80	0.51	0.72	0.32

From these data, one must conclude that in the non-sensitive species the wall is acting as a barrier, thus preventing the adsorption of the antibacterial to the underlying sensitive membrane.

From earlier results (Hamilton, 1968) with the uptake of TCS by *Staph. aureus* and *E. coli* it was apparent that the wall of the Gram negative organism was acting as a non-adsorbing barrier, and that consequently whole cells adsorbed considerably less antibacterial than did *E. coli* spheroplasts or whole cells of *Staph. aureus*. Figures for the uptake are given in Table 4. We suggested that this might be the general mechanism of non-sensitivity, but further data with TCC and CTAB uptake by *Staph. aureus*, *E. coli* and *Ps. aeruginosa* show that whole cells of non-sensitive species take up as much antibacterial as do spheroplasts or whole cells of sensitive species. We assume that in these cases the

Table 4. Uptake of antibacterials (M/B) by whole cells (C) and spheroplasts (S)

	TCS (0.5 μg/ml)		TCC (1.32 μg/ml)		CTAB (24 μg/ml)	
	C	S	C	S	C	S
Staph. aureus	0.75×10^5	—	3.6×10^5	—	6.3×10^6	—
E. coli	0.08×10^5	0.55×10^5	3.7×10^5	4.0×10^5	6.0×10^6	—
Ps. aeruginosa	—	—	2.8×10^5	2.1×10^5	4.6×10^6	2.5×10^6

Gram negative cell wall acts not only as a barrier preventing adsorption to the membrane, but as a barrier which is itself adsorbing.

The second aspect of this work which we hope to develop is the direct study of ion fluxes, oxidative phosphorylation and uncoupling in a cytoplasm—and wall-free membrane preparation from *E. coli*, such as is shown in Fig. 7. These

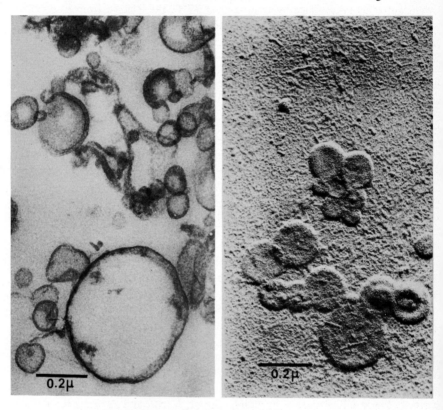

Figure 7. Membrane vesicles prepared from *E. coli*.

are positively stained sections and shadow cast specimens of membrane vesicles prepared by the method of Kaback and Stadtman (1966) and shown by these authors and by ourselves to be capable of the active transport of sugars and amino acids. It is hoped that this will be a useful system for studies of oxidative phosphorylation in bacteria and for a critical appraisal of the chemiosmotic hypothesis (Mitchell, 1966), using these membrane-active antibacterials as an experimental tool.

REFERENCES

Chappell, J. B. and Crofts, A. R. (1965). *Biochem, J.* **95**, 393.

Chappell, J. B. and Haarhoff, K. N. (1966). *In* "Biochemistry of Mitochondria" (E. C. Slater, Z. Kaniuga and L. Wojtczak, eds.), pp. 75-91. Academic Press, London and New York.
Hamilton, W. A. (1968). *J. gen. Microbiol.* **50**, 441.
Harold, F. M. and Baarda, J. R. (1967). *J. Bact.* **94**, 53.
Harold, F. M. and Baarda, J. R. (1968). *J. Bact.* **96**, 2025.
Höfer, M. and Pressman, B. C. (1966). *Biochemistry* **5**, 3919.
Hopfer, V., Lehninger, A. L. and Thompson, T. E. (1968). *Proc. natn. Acad. Sci. U.S.A.* **59**, 484.
Hotchkiss, R. D. (1944). *Adv. Enzymol.* **4**, 153.
Kaback, H. R. and Stadtman, E. R. (1966). *Proc. natn. Acad. Sci. U.S.A.* **55**, 920.
Mitchell, P. (1966). *Biol. Rev.* **41**, 445.
Mitchell, P. and Moyle, J. (1967). *Biochem. J.* **104**, 588.
Moore, C. and Pressman, B. C. (1964). *Biochem. biophys. Res. Commun.* **15**, 562.
Salton, M. R. J. (1951). *J. gen. Microbiol.* **5**, 391.

Introduction

H. N. CHRISTENSEN

*Department of Biological Chemistry,
The University of Michigan, Ann Arbor, U.S.A.*

We will not find it difficult, I think, to perceive the focus of the interest of Prof. Wilbrandt and his committee in the design of this program. The stage is to be set by a consideration by Prof. Katchalsky* of the irreversible thermodynamics of active transport. The only thing I can offer in this area is to pose a biological question: "How often is the supposedly irreversible movement of a component in the direction of its downward gradient of chemical potential actually made thermodynamically reversible by associated migrations, or by an associated change in the membrane?" Dr. Crane's* remarks will, I judge, concern cases that may be removed from the area of irreversible thermodynamics because of the linkage of fluxes. What is especially exciting is that Crane will apparently present evidence that the site by which Na^+ is brought into the linkage has its orientation changed by insulin and that it is perhaps the site of the primary Na^+-pump.

It is interesting that the several papers, after the theoretical one by Prof. Katchalsky, have been selected from the area of sugar transport. It is also interesting how adequate this choice proves for an exploration of some of the major questions in the area. Of these questions, the one that obviously played the largest role in the selection of the program was: "Does an enzyme molecule present the reactive site described by the kinetics of transport, or does the transport site lack the destabilizing activity which is ordinarily taken to define an enzyme?" I hope I have stated this question so that it stands as a significant conceptual matter, and not as a quibble about what we mean by the word *enzyme*.

The remaining four papers, beyond the two I have mentioned, have this question as an important concern. Doctors Gachelin and Kepes will evaluate a number of associations between the transport of α-methylglucoside into *E. coli*, and its phosphorylation by an enzymatic system. By very careful consideration, they are, I judge, tending to conclude that a step unique to transport precedes the phosphorylation.

* The papers by Katchalsky and Crane are not included in this volume.

Other sessions elsewhere have occasionally been heavily loaded in favor of what we may call the enzymologist's intuition about transport, or else in favor of the narrowest notion of what we mean by a transport carrier. Neither of these two all-or-none biases on our primary question may prove fortunate; different transport systems may well coexist, and even interact: catalyzed in the one case by sites on protein molecules that produce no destabilization of the substrate, and in the other case by sites that catalyze a persisting change in the covalent structure of the substrate. Perhaps these two may even occur sequentially in transport, as we are perhaps to hear for the present case.

Doctor Kotyk will consider the question of what reactivities we should look for on the separated transport protein. His abstract suggested that we would do best to look for stoichiometric binding activity for the substrate, and for immunological reactivity associated with the transport system.

Doctor Semenza has pursued a fascinating association between a Na^+- and sugar-binding component of the intestinal monosaccharide transport system, and a sucrose- and isomaltose-hydrolyzing enzyme. Both of these activities occur together in an artificial solution obtained after papain digestion; an immunologic activity identified with the transport system is also retained by the solubilized preparation. We shall see whether his findings establish his view that sucrase, nevertheless, is not the carrier.

Doctor Alvarado considers the possibility that the site at which phlorizin binds at the intestinal wall to inhibit sugar transport may be the one recently found to be hydrolytic to this glucoside. He tends towards the conclusion that this cleavage is not *necessary* to the inhibition of sugar absorption; but we must wait to hear whether he considers the site, nevertheless, to have an *incidental* hydrolytic activity.

The question of how enzymatic sites are related to transport occurs also in Crane's paper; but there I believe it is ATP and not the sugar which is presumed to bind at a site that catalyzes hydrolysis or group transfer.

I note in this Symposium an attractive intermingling of papers on *transport into cells* and *transport across cells.* One can only deplore the occasional tendency to treat these subjects, historically or conceptually, as unrelated fields.

Perhaps I may take it as an appropriate role to speak briefly of two aspects in which results with non-sugar substrates may supplement those using sugars—and I refer of course to amino acids. The first possible advantage is that the range of amino acid molecules that one can make available can provide certain *sensor groups* to indicate what is going on in the vicinity of the transport site. When we proposed in the interval 1952-58 that one or both of the ordinary alkali-metal ions binds to the same carrier as the amino acid for transport, we had almost no basis for guessing what position Na^+ or K^+ takes with respect to the position taken by the amino acid. But as illustrated in Fig. 1, we have now discovered a similar reaction in amino acid transport, in which the combination of NA^+ and

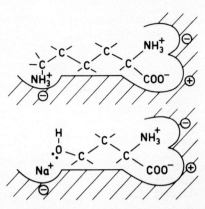

Figure 1. Scheme to suggest how a neutral amino acid plus sugar reacts with the cationic amino acid transport site. The Na^+-dependent reactivity of the amino acid is optimal, both in the inhibition of cationic amino acid uptake and in stimulating the exodus of basic amino acids when the chain has five linear carbon atoms, and when an oxygen or sulfur atom is exposed on carbon 5 (Christensen and Handlogten, 1969; Christensen et al., 1970).

neutral amino acid competes for transport with, and exchanges across the plasma membrane for, a cationic amino acid molecule. As I have indicated in detail elsewhere (Christensen and Handlogten, 1969; Christensen et al., 1969), the Na^+ undoubtedly takes the position otherwise occupied by the terminal amino group. We have distinct indications of the distance between this position and the points at which the α-amino and carboxyl groups are fixed; appropriately positioned oxygen or sulfur atoms assist the reaction with Na^+. Perhaps sugars may offer a similar possibility, our attention falling naturally on the oxygen at c-2.

As a second advantage, amino acid molecules can be built with very great resistance to modification by enzymes. The same may or may not be possible with sugar substrates. I refer not merely to non-metabolizable substrates, which are of course known for both the amino acid arena and the sugar satrapy, but to substrates resistant even to structural distortion. One of my colleagues asked me a year ago, "How do you know that the binding site is not activating and distorting the molecule of cycloleucine or the molecule of α-aminoisobutyric acid? The mere circumstances that these molecules prove strong enough to survive such influences does not mean that the distortion was unnecessary to transport".

We did not design the amino acid molecule shown in Table 1 for that purpose, but instead to obtain a specific substrate for transport by System *L*. You will note we accomplished that purpose; one of the isomeric forms of this bicylic amino acid is transported in a perfectly typical way by an *L*-type system in each of the several species studied, and, as far as we can tell, by no other system. Note that the same isomer also stimulates the release of insulin in the rat

Table 1. Biological reactivities identified for the conformational isomers of 2-aminobicyclo(2,2,1)heptane-2-carboxylic acid.

Function	COO⁻ / NH₃⁺	NH₃⁺ / COO⁻
Influx into *E. coli*	rapid	0
Influx into pigeon r.b.c.	+++	+
Influx into human r.b.c.	+++	+
Influx into Ehrlich cell	+++++	+
Stimulation of insulin release	strong	0
Binding to leucine-binding protein of *E. coli*	+++++	+
Metabolism in the rat	0	0
Insulin stim. of uptake by rat diaphragm	0	0

Features: Unmetabolizable; atoms presented in rigid array, subject to minimal distortion. The *endo* and *exo* conformations have been assigned arbitrarily above. Quantitative comparisons of the functions appear elsewhere (Christensen and Cullen, 1969; Christensen *et al.*, 1969).

in much the same way that leucine does. But what I want to emphasize at this time is the *inflexibility* of this molecule. Atoms that correspond to those of leucine take defined positions in space. Examination of a space-filling model shows that the structure can scarcely undergo distortion during transport. All that remains is to wonder whether the binding site applies some abortive stresses to this substrate molecule. The possibilities for extension of the use of sensor groups and rigid structures appears very substantial to us.

The leucine-binding protein associated with amino acid transport in *E. coli* makes the same choice among the isomers of this model amino acid: that isomer shows a much larger inhibition of leucine binding than does the other isomer, to strengthen the evidence for the role of this protein.

The questions raised in this Symposium have already proved capable of arousing emotions—I think they will do so on the western shores of the Atlantic—so I believe we are likely to have some important discussion.

Since it was for Prof. Wilbrandt and his committee that the speakers agreed to come and present their results and thoughts, rather than for me, I have the advantage that I can take a harder line than might otherwise seem courteous to encourage the speakers to preserve time for discussion by staying closely within their scheduled intervals.

REFERENCES

Christensen, H. N. and Cullen, A. M. (1969). *J. biol. Chem.* **244**, 1521.
Christensen, H. N. and Handlogten, M. E. (1969). *FEBS Lett.* **3**, 14.

Christensen, H. N., Handlogten, M. E., Lam, I., Tager, H. S. and Zand, R. (1969). *J. biol. Chem.* **244**, 1510.
Christensen, H. N., Handlogten, M. E. and Thomas, E. L. (1969). *Proc. natn. Acad. Sci. U.S.A.* **63**, 948.

Phosphorylative Transport of Sugars in *E. coli*

G. GACHELIN and A. KEPES

*Service de Biochimie Cellulaire, Institut Pasteur, and
Laboratoire de Biologie Moléculaire, Collège de France, Paris, France*

The discovery of bacterial permeases (Rickenberg *et al.*, 1956) was the first breakthrough in attributing to specific proteins an essential role in the transport of a number of metabolites through bacterial membranes. Permeases have been defined (Cohen and Monod, 1957) as the key proteins specialized in transport, and specific for a given substrate or family of substrates. Their specificity was ascertained, besides the narrow spectrum of substrate molecule configuration, by the genetic identification and very often by the specific regulatory mechanism of their biosynthesis. Specialization in transport was defined as the absence of a biochemical transformation of the transported substrate which could be a first step in metabolism. In a large number of instances the unchanged intracellular substrate was identified as the exclusive or predominant product of transport. Nevertheless, it was assumed that a transient intermediate confined in the thickness of the bacterial envelope could exist (Kepes, 1960), but that it would be transformed back into the native substrate when it reached the aqueous medium of the cytoplasm. Since the transport was usually found to be thermodynamically active, the intervention of metabolic energy in the form of a high energy linkage was admitted, and high energy donors being mostly phosphoric anhydrides or esters, phosphorylated intermediates have been investigated. Such phosphorylated intermediates have been detected in *E. coli* as a result of the transport of αmethylglucoside (αMG) (Rogers *et al.*, 1962) a "non-metabolizable" analog of glucose (Cohen and Monod, 1957; Hoffee *et al.*, 1964), and in the case of most sugars transported by *S. aureus* (Anderson *et al.*, 1968; Eagan and Morse, 1965; Hengstenberg *et al.*, 1967).

These phosphorylated intermediates and their implication in the mechanism of transport attracted much more attention after the discovery (Kundig *et al.*, 1964, 1965; Simoni *et al.*, 1967) of a phosphoenolpyruvate-hexose phosphotransferase system catalyzing the following two reactions:

$$\text{Hpr} + \text{PEP} \xrightleftharpoons[\text{Mg}^{2+}]{\text{Enz I}} \text{P} - \text{Hpr} + \text{Pyruvate}$$

$$\text{P} - \text{Hpr} + \text{hexose} \xrightarrow[\text{Mg}^{2+}]{\text{Enz II}} \text{hexose-6-P} + \text{Hpr}$$

where Hpr is a low molecular weight, heat stable protein, phosphorylated on a histidyl residue. Enzyme I is in the soluble cell extract, and enzyme II is particulate, presumably membrane bound. It is supposed that a different enzyme II is used for different sugars on the basis that their relative rate of phosphorylation varies with the carbon source used for growth.

In contrast, the two soluble protein components enzyme I and Hpr are common to a number of phosphotransferase systems and their loss by mutation causes multiple deficiencies in carbohydrate uptake (Simoni et al., 1967).

The system may include a fourth protein component, the K_m factor (Simoni et al., 1968), which increases the affinity of enzyme II for a given substrate.

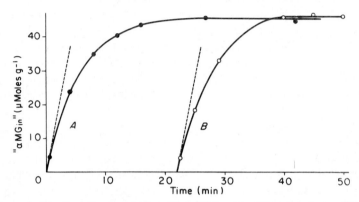

Figure 1. Kinetics of α-methyl glucoside (αMG) uptake (A) and exchange (B) in E. coli 3300. Bacteria in exponential growth on mineral medium 63 with 4 g/litre glycerol as carbon source were stopped by 50 μg/ml chloramphenicol at a cell density of 300 μg wet weight/ml. After equilibration at 25°C, [^{14}C] αMG was added to make the final concentration 2×10^{-4} M in experiment A. Samples were filtered on millipore membranes of 0.45μ pore diameter (Kepes, 1960). In experiment B non-radioactive αMG was added at zero time and tracer amounts of [^{14}C] αMG at 22 min.

Thus the PEP-hexose phosphotransferase includes at least three protein components, one of which is membrane bound and hexose specific. The sugar phosphate which results from the action of phosphotransferase might be the first metabolic intermediate or might be dephosphorylated and possibly phosphorylated again by an ATP-dependent kinase. This duplication of systems yielding the same product was another argument in favor of the distinct role of PEP-hexose phosphotransferase in sugar uptake (Simoni et al., 1967; Tanaka and Lin, 1967), but it contradicts the original definition of a permease as distinct from any metabolic enzyme.

The αMG permease in E. coli behaves very similarly to other permeases, e.g. βgalactoside permease, when observed kinetically with the millipore filtration technique. As shown in Fig. 1, the uptake of this sugar starts immediately upon its addition to the bacterial suspension, it slows down after a few minutes and reaches a plateau at about 15 min. The shape of the curve closely approximates an exponential function. When the plateau is reached, intracellular αMG exchanges with αMG in the medium, as seen from curve B of Fig. 1. Curve B is obtained when cells, previously loaded with non-radioactive substrate in conditions identical to those used for sample A, are supplemented with tracer amounts of radioactive αMG at 22 min. Initial rate of exchange is identical to initial rate of uptake and after some 15 min the totality of intracellular αMG is exchanged. This means that the accumulated αMG behaves as a single pool, whatever derivative of αMG is present is either in negligible amount or in fast equilibrium with the free sugar. The steady state reflects the balance between rate of uptake and rate of exit.

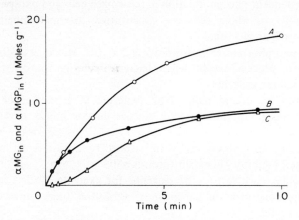

Figure 2. Experiment similar to that described in Fig. 1A, except that besides millipore filtration (A) samples have been collected in nine volumes of 90 per cent ethanol containing 11 mg/ml $BaBr_2$ to precipitate αMG-6-P, which was filtered and counted (B). Free intracellular sugar (C) is independently determined (Gachelin, 1970).

Nevertheless, as shown in Fig. 2, intracellular "αMG" (A) is composed of approximately equal amounts of αMG-6P (B) and free sugar (C). The kinetics of these two pools are different. The initial rate of appearance of αMG-6P is virtually the same as the rate of total uptake and free sugar appears after a definite lag, suggesting a precursor-product relationship between phosphorylated and free αMG. It is noteworthy that in spite of the permanent turnover of these two pools, no phosphorylated αMG appears in the medium. This fact inevitably leads to the postulation of a hexose phosphohydrolase.

In order to establish a close correlation between PEP hexose phosphotransferase and αMG transport and to show that it was the rate limiting step of total αMG uptake, it was of interest to have a simple and accurate measurement without necessarily separating all constituents and reconstructing the system. This possibility was provided by a simple toluene treatment of the cells

(Gachelin, 1969). This treatment abolishes the permeability barrier of the cell membrane and leads to a leak of the intracellular pool of metabolites, without causing immediately gross structural alterations and a heavy loss of intracellular proteins. The toluene treated cells do not accumulate radioactive αMG or αMG-6P and they do not phosphorylate the sugar, unless PEP is added to the medium. In the presence of PEP, phosphorylated αMG appears linearly in the medium with no tendency to plateau and no evidence of dephosphorylation. PEP cannot be replaced by ATP or a number of high or low energy phosphate esters.

The maximal velocity of PEP-dependent phosphorylation in toluenized bacteria closely follows the maximal velocity of uptake *in vivo*.

Both PEP-dependent phosphorylation in toluenized cells and phosphorylating transport of αMG in whole cells are constitutive; maximal velocity is not modified by growth on glucose.

The K_m of phosphotransferase for αMG is 2×10^{-4} M, which is also the K_m of the transport process except when bacteria were grown or glucose or submitted to uncoupling agents.

Both activities are found reduced or abolished in a number of mutants selected, for example, for the character glucose diauxie minus, or no growth on glucose.

Both have the same specificity with respect to a variety of sugars, as estimated by their inhibitory constant towards αMG.

Both are highly sensitive to −SH reagents, for example *N*-ethyl maleimide (NEM), and the concentration dependence, as well as time course of inactivation, are very similar.

The close correlation between the PEP-dependent phosphorylation and trans-membrane transport of αMG is even more strikingly demonstrated by the work on protoplast ghost membranes by Kaback (1968). These membranes, which form closed vesicles, are able to accumulate αMG in the phosphorylated form. The process is different from *in vivo* transport in that it is dependent on added PEP and it does not exhibit turnover or dephosphorylation. When such vesicles are loaded passively with [^{3}H] glucose in the absence of PEP and then resuspended in a buffer containing PEP and [^{14}C] glucose, [^{14}C] glucose-6-phosphate is synthesized and appears in the vesicles, whereas [^{3}H] glucose-6-phosphate appears in negligible amounts and only after a lag, suggesting that the [^{3}H] glucose had first to leak out before its phosphorylating accumulation could take place.

In spite of these observations, which lend strong support to the essential role of PEP-hexose phosphotransferase in transport, the detailed mechanism of the latter is not easy to construct. Is Hpr located on the outer side of the membrane, between membrane and cell wall, as inferred from its release by a "cold shock" (Kundig *et al.*, 1966; Neu and Heppel, 1965)? Then enzyme I and PEP should also be located in this compartment and the hexose should be phosphorylated before crossing the membrane, with Hpr phosphorylated inside. One should admit either that P-Hpr is crossing the membrane, transferring the phosphate radical to hexose via enzyme II and crossing back to be rephosphorylated or the whole sequence of the two reactions is taking place in the inner compartment. In the last case enzyme II should first play the role of an αMG-carrier before it plays the role of phosphotransferase, while in the first hypothesis it is a carrier

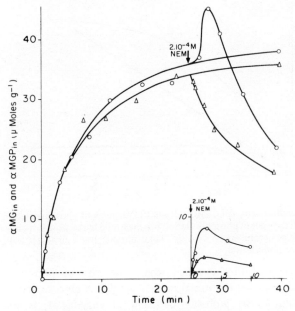

Figure 3. Effect of NEM on αMG uptake and phosphorylation. Bacteria, concentration of αMG and techniques are the same as in Figs. 1 and 2. NEM (2×10^{-4} M) was added at 25 min (main graph) or at zero time (insert). —o—o— Total intracellular radioactivity; —△—△— barium precipitable radioactivity. – – – External concentration.

of its product αMG-P. Whatever the case might be, the complete *in vivo* transport (uptake, exchange, exit), necessitates at least one more protein component, a phosphohydrolase or phosphotransferase, if the phosphate moiety of hexose-6-P is transferred to another acceptor. This postulated enzyme could not be demonstrated until now, either in toluenized bacteria or in membrane vesicles.

In contrast, a number of facts suggest a step of transport distinct from the action of phosphotransferase. As shown in Fig. 3, addition of NEM to bacteria, which are at the steady state of accumulation of αMG, with a high proportion of intracellular αMG-6-P, causes an immediate decrease of the concentration of sugar phosphate, while free sugar first increases and its leak is delayed by more

4*

than 3 min. With the same bacteria, NEM added at the same time as [^{14}C] αMG decreases phosphorylation to a much greater extent than sugar accumulation (Gachelin, 1970). Furthermore, when glucose is added a few minutes after NEM, when free αMG is high and αMG-6-P already low, the exit of αMG is strongly accelerated. This phenomenon is strongly reminiscent of a counterflux interpreted in terms of a mobile carrier operating in spite of the well advanced inactivation of the phosphotransferase system by NEM.

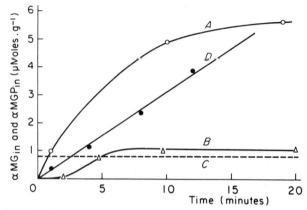

Figure 4. Accumulation of αMG, αMG-6-P and PEP-dependent phosphorylation of αMG in *E. coli* 3300-I. PEP-dependent phosphorylation was measured in the same suspension as uptake after a short toluene treatment. A —o—o— Total uptake; B —△—△— αMG-6-P; D —●—●— PEP-dependent phosphorylation.

Figure 4 shows another instance where the rate of phosphorylation is inadequate to account for the initial rate of uptake. A glucose negative mutant 3300-I (Gachelin, 1970) probably deficient in enzyme II, is still able to accumulate free αMG, while αMG-6-P appears in abnormally low proportion and only after several minutes.

Galactose competes for αMG uptake but is not phosphorylated by toluenized bacteria in the presence of PEP (Gachelin, 1970), and does not inhibit competitively the phosphorylation of αMG in the same system.

A mutant, lacking enzyme I or Hpr, when previously equilibrated with glucose shows a biphasic curve of exchange (Fig. 5), whereas the simultaneous addition of galactose reconstitutes a classical pattern of equilibration of αMG (Gachelin, 1970). Here again the overshoot can be interpreted in terms of counterflux of glucose against αMG, which is inhibited by galactose.

Before all these facts can be incorporated into a model, a more detailed genetic analysis is necessary, concerning mainly the uniqueness of the glucose transport system or the possibility of transport to various extents by related transport systems.

A paradoxical observation about αMG transport is the absence of inhibitory

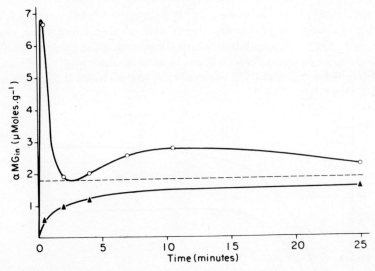

Figure 5. Exchange of αMG in *E. coli* 3300 G 10, grown on L. medium, after preload with non-radioactive glucose, 5×10^{-3} M.A.: αMG 4×10^{-4} M.B: αMG 4×10^{-4} M, galactose 5×10^{-3} M.

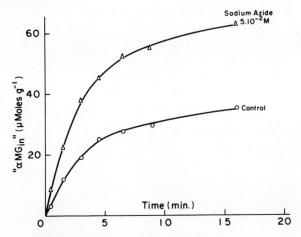

Figure 6. Stimulation of αMG uptake by Na_3N in *E. coli* 3300 grown on glycerol. αMG, 10^{-4} M; Na_3N, 5×10^{-2} M.

effect of uncoupling agents like Na azide and 2,4-dinitrophenol (Cohen and Monod, 1957) and even its stimulation by these inhibitors (Englesberg *et al.*, 1961) (Fig. 6). This stimulation is due to a shift of the apparent K_m of the system for αMG (Englesberg *et al.*, 1961) towards lower values, and is not observable in *E. coli* grown on glucose as the sole carbon source, which exhibits the lower apparent K_m without inhibitor (Kepes, 1964). It was found that Na_3N

does not modify the properties of PEP-dependent phosphorylation by toluenized bacteria, the passive diffusion of αMG across the membrane, or the rate of exit. The increase in intracellular radioactivity is due predominantly or exclusively to an increase of αMG-6-P, and in some instances when no increase in total intracellular radioactivity is observed, αMG-6-P increases at the expense of free αMG (Fig. 7).

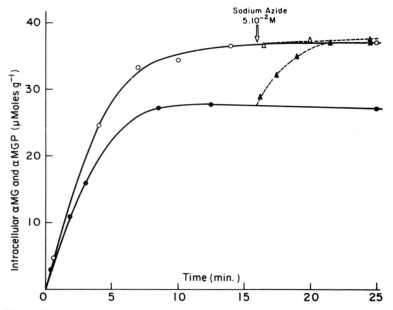

Figure 7. Stimulation of αMGP production in *E. coli* grown on medium 63 glycerol. αMG, 10^{-4} M; —o—o— uptake; —△—o— uptake in presence of 5×10^{-2} M Na$_3$N; —●—●— αMGP; —▲—▲— αMGP in presence of 5×10^{-2} M Na$_3$N.

Among the many modifications of intracellular metabolites caused by uncoupling agents, the increase of PEP was likely to be the most significant in view of the requirements of the phosphotransferase.

Direct measurement of the intracellular pool of PEP and of pyruvate gave the results summarized in Table 1. Thus it appears that the production of PEP and also its utilization increases in the presence of azide and it is difficult to ascertain at the present stage that the increase of αMG uptake is caused by the increased production of PEP. Nevertheless the effect of sodium fluoride supports this view. This inhibitor, at a concentration of 5×10^{-2} M does not dramatically inhibit αMG uptake and phosphorylation in aerobic bacteria, but as shown in Fig. 8, it completely abolishes the stimulatory effect of Na azide.

The generalization of the PEP-phosphotransferase-dependent transport should await a more extensive study than presently available. It seems to be operating in *E. coli* (Fraenkel *et al.*, 1964; Tanaka *et al.*, 1967), in *B. subtilis* (Lepesant and

Table 1. Variation of the intracellular pool of PEP and pyruvate in the presence of αMG and of Na azide in μmoles/gram dry weight.

μmoles/g^{-1}	Control	αMG (10^{-4} M)	Na$_3$N (5×10^{-2} M)	αMG (10^{-4} M) + Na$_3$N (5×10^{-2} M)
PEP	1.24	0.38	0.40	0
Pyruvate acid	0.11	0.45	1.98	2.37

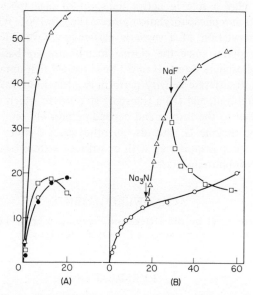

Figure 8. Combined effect of NaF and Na$_3$N on the uptake of αMG. (A) Inhibitors added 2 min before αMG; (B) inhibitors added in sequence as indicated by the arrows. —o—o— Control; —●—●— NaF 2×10^{-2} M; —△—△— Na$_3$N 2×10^{-2} M; —□—□— NaF + Na$_3$N 2×10^{-2} M each.

Dedonder, 1968), in *Aerobacter* (Tanaka and Lin, 1967), in *Salmonella* (Simoni et al., 1967) and in *Staphylococcus* (Eagan and Morse, 1965). The list of the relevant sugars might be established by the finding of a phosphorylated pool or by the simultaneous loss of their transport systems following a single mutation in the non-specific part of phosphotransferase, either enzyme I or Hpr. Until now, the following sugars have been found to use this pathway in one or several species of bacteria: glucose, fructose, mannose, mannitol, melibiose, sorbitol, glycerol, *N*-acetyl glucosamine, *N*-acetyl mannosamine, maltose (Simoni et al., 1967; Tanaka and Lin, 1967; Tanaka et al., 1967). Concerning the transport of β-galactosides in *E. coli* by β-galactoside permease, the data are conflicting.

Phosphorylation of lactose, but not of TMG or TDG, has been observed (W. Kundig, personal communication) and the uptake of TMG has been inactivated by cold shock and reactivated by addition of excess Hpr (Kundig et al., 1966). Nevertheless, the pleiotropic mutants deficient in the uptake of several sugars have an unimpaired galactoside permease activity.

It is altogether unlikely that all sugars and sugar derivatives follow the PEP phosphotransferase pathway for their transport. Pleiotropic mutants show, besides a deficiency for the utilization of a number of sugars, a normal utilization of a number of carbohydrates. Moreover, specific transport systems are known to play a role in active transport of sugar phosphates without a dephosphorylation-rephosphorylation process (Pogell, 1966; Winkler, 1966).

The phosphorylation of a sugar is particularly favorable to its retention within the cytoplasm since the plasma membranes are basically very poorly permeable to phosphate esters. It might be suggested that uptake systems linked to PEP phosphotransferase activity perform two functions, (a) phosphorylation, essential for retention, and (b) a transport step *per se* which might be more or less closely linked to the former and depending possibly on the same molecular species, namely enzyme II. But also in other cases distinct carrier molecules might perform the transport step with or without a close relation in space and time with the phosphotransferase.

ACKNOWLEDGEMENTS

This work is supported by the Délégation Générale à la Recherche Scientifique et Technique and the Commissariat à l'Energie Atomique.

REFERENCES

Anderson, B., Kundig, W., Simoni, R. D. and Roseman, S. (1968). *Fedn Proc. Fedn Am. Socs exp. Biol.* **27**, 643.
Cohen, G. N. and Monod, J. (1957). *Bact. Rev.* **21**, 169.
Eagan, J. B. and Morse, M. L. (1965). *Biochim biophys. Acta* **97**, 310.
Englesberg, E., Watson, J. A. and Hoffee, P. A. (1961). *Cold Spring Harb. Symp. quant. Biol.* **26**, 261.
Fraenkel, D. G., Falcoz-Kelly, F. and Horecker, B. L. (1964). *Proc. natn. Acad. Sci. U.S.A.* **52**, 1207.
Gachelin, G. (1969). *Biochem. biophys. Res. Commun.* **34**, 382.
Gachelin, G. (1970). (Manuscript in preparation).
Hanson, T. E. and Anderson, R. L. (1968). *Proc. natn. Acad. Sci. U.S.A.* **61**, 269.
Hengstenberg, W., Eagan, J. B. and Morse, M. L. (1967). *Proc. natn. Acad. Sci. U.S.A.* **58**, 274.
Hoffee, P., Englesberg, E. and Lamy, F. (1964). *Biochim. biophys. Acta* **79**, 337.
Kaback, H. R. (1968). *J. biol. Chem.* **243**, 3711.
Kepes, A. (1960). *Biochim. biophys. Acta* **40**, 70.

Kepes, A. (1964). *In* "Cellular Functions of Membrane Transport", p. 155. Prentice-Hall, Englewood Cliffs, N.J.
Kundig, W., Ghosh, S. and Roseman, S. (1964). *Proc. natn. Acad. Sci. U.S.A.* **52**, 1067.
Kundig, W., Kundig, F. D., Anderson, B. E. and Roseman, S. (1965). *Fedn Proc. Fedn Am. Socs exp. Biol.* **24**, 658.
Kundig, W., Kundig, F. D., Anderson, D. and Roseman, S. (1966). *J. biol. Chem.* **241**, 3243.
Lepesant, J. A. and Dedonder, R. (1968). *C. r. hebd Seanc. Acad. Sci., Paris.* **267**, 1109.
Neu, M. C. and Heppel, L. A. (1965). *J. biol. Chem.* **240**, 3685.
Pogell, B. M., Maity, B. R., Frumkin, S. and Shapiro, S. (1966). *Archs Biochem Biophys.* **116**, 406.
Rickenberg, H. W., Cohen, G. N., Buttin, G. and Monod, J. (1965). *Annls Inst. Pasteur, Paris* **91**, 829.
Rogers, D. and Yu, S. M. (1962). *J. Bacteriol.* **84**, 877.
Simoni, R. D., Levinthal, M., Kundig, F. D., Kundig, W., Anderson, B., Hartman, P. E. and Roseman, S. (1967). *Proc. natn. Acad. Sci. U.S.A.* **58**, 1963.
Simoni, R. D., Smith, M. F. and Roseman, S. (1968). *Biochem. biophys. Res. Commun.* **31**, 804.
Tanaka, S. and Lin, E. C. C. (1967). *Proc. natn. Acad. Sci. U.S.A.* **57**, 913.
Tanaka, S., Fraenkel, D. G. and Lin, E. C. C. (1967). *Biochem. biophys. Res. Commun.* **27**, 63.
Winkler, H. H. (1966). *Biochim. biophys. Acta* **117**, 231.

Present Situation in the Identification of Membrane Transport Proteins in Single Cells

A. KOTYK

Laboratory for Cell Membrane Transport,
Institute of Microbiology, Czechoslovak Academy of Sciences,
Prague, Czechoslovakia

It is perhaps noteworthy that the present concerted attack on problems connected with the isolation of membrane carriers was begun almost simultaneously about five years ago in at least four laboratories in Europe and America. One can wonder about the reasons for this rather late start since the technical equipment required for the isolation attempts is not particularly complicated and had been available for perhaps ten years. It can apparently be ascribed to the level of sophistication attained by the investigators and to their awareness of the futility of further purely kinetic research of transport processes, although it was this very approach which proved beyond doubt the existence of membrane systems which are specifically involved in the transport of molecules into cells and which thus might be amenable to a molecular-level purification.

Three basic approaches have been used in the work on erythrocytes and micro-organisms, and these are briefly described below.

I. TECHNIQUES USED FOR ISOLATION OF MEMBRANE CARRIERS

1. Double labelling of inducible carriers

It was as a result of research in Stein's laboratory that a dual labelling was applied to an inducible transport system, viz. the β-galactoside permease of *E. coli* (Kolber and Stein, 1966). A bacterial culture was grown in the presence of the transport inducer (thiomethyl β-galactoside) and ^{14}C-labelled phenylalanine; another culture was grown without inducer in the presence of ^{3}H-labelled phenylalanine. The cultures were combined and analysed for unequally ^{3}H + ^{14}C-labelled proteins. The particle-free supernatant was found to contain three peaks of enrichment with ^{14}C, one corresponding to β-galactosidase, another to thiogalactoside transacetylase, and the third possibly to a β-galactoside permease. It was somewhat surprising that the "permease" material should have been found in the soluble fraction rather than in the plasma membrane, but it followed from subsequent experiments carried out by Kolber in Prague (Kolber and Stein, 1967) that ^{14}C-enriched material was likewise present in proteins solubilized from membranes of a galactosidase-less mutant of *E. coli*. The protein found previously in the soluble fraction may either have been washed out from the membranes during the fractionation procedure or else it represents a component (subunit) of the membrane-bound material.

The double-labelling procedure was applied in our Prague laboratory to the isolation of the inducible galactose carrier from baker's yeast (Haškovec and Kotyk, 1969), and a peak presumed to correspond to the carrier was found in the elution diagrams of solubilized membrane proteins from both the wild strain and the galactokinase-less mutant of *Saccharomyces cerevisiae* (Fig. 1).

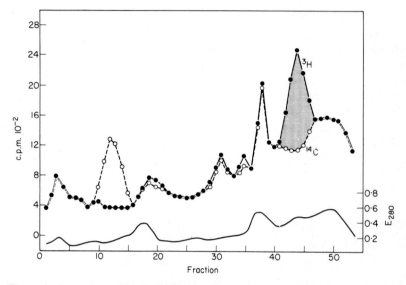

Figure 1. Separation of NaI-solubilized membrane proteins of baker's yeast galactokinase-less mutant, doubly labelled with ^{14}C (uninduced) and ^{3}H (induced). Column of DEAE-cellulose (1.6 x 30 cm) eluted with a linear (0.005-0.5M) gradient of NH_4HCO_3. Two ml fractions collected. Their protein content estimated at 280 nm (lower curve) and their radioactivity due to ^{3}H and ^{14}C counted in a liquid-scintillation counter. The hatched area is assumed to correspond to the induced galactose-transporting protein.

2. Inhibitor binding to membrane carriers

Another of the contributions of Stein's laboratory to the investigation of carrier proteins was the application of the finding that dinitrofluorobenzene inhibits the glucose transport by human erythrocytes in proportion to the square of its concentration which indicates more than one binding site for this inhibitor at the transport component. By treating erythrocytes with 10 mM ^{14}C-labelled

DNFB for 1 min and with 1 mM ^{3}H-labelled DNFB for 10 min, mixing the cells and analysing the mixture, he obtained an unequally labelled protein with DNP-cysteine and another reactive residue which may have been identical with the glucose transport system (Stein, 1964). Pursuing this line further, with Bobinski, Bonsall and Hunt as collaborators (Bobinski and Stein, 1966; Bonsall and Hunt, 1966), he obtained clear indications of a glucose-binding protein in erythrocyte ghosts and their solubilized fractions.

The technique of specific labelling of the transport system (or of its part) was used independently by Fox and Kennedy and co-workers (Fox and Kennedy, 1965; Fox et al., 1967) who devised an elegant method based on blocking the transport system for galactosides in *E. coli* with *N*-ethylmaleimide (NEM). They first treated their bacteria (induced and uninduced) with unlabelled NEM in the presence of thiodigalactoside in order to protect the permease binding sites against the attack of NEM. Subsequently, the cells were washed and treated with [^{14}C]NEM (induced cells) and with [^{3}H]NEM (uninduced cells). After combining and analysing the culture for uneven labelling with the two isotopes they isolated a component which they termed M-protein and which is apparently involved in the uptake of β-galactosides by the *E. coli* cells. The protein has since been isolated in a pure state and its molecular weight determined (Jones and Kennedy, 1968).

3. Affinity of membrane-bound proteins for their substrates

It is ironic that the real breakthrough in the isolation and complete purification of membrane carriers has been achieved without auxiliary techniques by simply following the binding affinity of various membrane components. It was A. B. Pardee of Princeton (Pardee, 1966, 1967; Pardee and Prestidge, 1966; Pardee *et al.*, 1966) who reached the goal first by isolating, in a crystalline form, a cell-membrane component of *Salmonella typhimurium* which is involved in the transport of sulphate by this bacterium and which is not present in repressed cells. Its isolation was aided by sonication and by osmotic lysis.

Treatment of bacteria with 0.5M sucrose (with some Tris and EDTA) and, subsequently, with a very dilute solution of $MgCl_2$ produces an osmotic shock which results in the release of various membrane components into the medium, among them parts of transport systems. This technique, described by Heppel's group (Heppel, 1967) was used in at least two other laboratories where crystalline or very pure transport proteins were isolated, viz. those of Oxender (Penrose *et al.*, 1968; Piperno and Oxender, 1966) and Anraku (1967, 1968). In the former, a transport protein for L-leucine, in the latter those for D-galactose and L-leucine, were isolated from *E. coli.*

In our Prague laboratory, some progress was made towards isolating the constitutive glucose carrier from baker's yeast, the most promising solubilization agent appearing to be M NaI.

At the time of writing this paper, work is in progress on the isolation of other transport proteins from *E. coli* cells, namely those for L-arabinose and glucose-6-phosphate, as well as those for L-arabinose and mannitol from *Pseudomonas natriegens*.

The present achievements in the isolation of carrier proteins are summarized in Table 1.

II. TESTS FOR TRANSPORT PROTEINS

Owing to the lack of any chemical reaction undergone by substrate upon its binding to the transport protein, and because the main characteristic of the process is its vectorial character, the only way of testing for the presence of a transport protein *in vitro* is to make use of its affinity for the transported substrate, taking care that there is no other protein in the reaction mixture besides the "carrier" that would show affinity for substrate. Two principal techniques are in use, both of them employing labelled substrates.

(a) Equilibrium dialysis, usually across a cellophane membrane (Visking tubing and the like), which is most straightforward as far as the interpretation of results is concerned but usually requires a considerable time to give useful data (danger of bacterial contamination and of breakdown of the protein in question).

The attainment of equilibrium binding can be considerably speeded up by pressure ultrafiltration, the only complicating factor of the procedure being the fact that the decrease of ligand concentration in the ultrafiltrate must be taken into account.

(b) "Column chromatography" by two alternatives: (i) The suspected carrier protein is adsorbed to a polymer support (Bobinski and Stein, 1966) (e.g. DEAE-cellulose) and the ligand in solution is passed through the column. The retardation of appearance of the ligand in the effluent (in comparison with a non-binding, differently labelled analogue, e.g. $[^3H]$D-glucose and $[^{14}C]$L-sorbose) yields semi-quantitative information on the affinity of the membrane material for its substrate. (ii) A column of Sephadex-type gel is equilibrated with a solution of the ligand (Fairclough and Fruton, 1966; Hummel and Dreyer, 1962) and the same solution, except that it also contains the protein, is run through the column. At the excluded volume of the column a peak of greater radioactivity emerges from the column, corresponding to the protein-substrate complex. This peak is followed by a trough of the same magnitude.

An unequivocal demonstration of the presence of a transport protein in a preparation can be provided by reconstituting from it (and a non-transporting membrane) a functional system. This has been done in the laboratories of Oxender, Anraku and Roseman where the osmotically-released transport protein was applied to transport-less, shocked cells, thus causing them to regain their

Table 1. Some properties of purified transport proteins

Transported substrate	Source of protein	Molecular weight	Crystals	K_{diss} (mM)	$K_{transport}$ (mM)	Ref.
Sulphate	S. typhimurium	32,000	+	0.0001	0.004	Pardee et al., 1966; Pardee, 1967
β-Galactoside	E. coli	31,000	—	?	0.6	Jones and Kennedy, 1968; Winkler and Wilson, 1966
L-Leucine	E. coli	36,000	+	0.001 (0.0005)	0.001	Piperno and Oxender, 1966; Penrose et al., 1968
L-Leucine	E. coli	36,000	+	0.002	0.001	Anraku, 1968
D-Galactose	E. coli	35,000	—	0.001	0.004 (0.0005)	Anraku, 1968; Rotman and Radojkovic, 1964
D-Glucose	S. cerevisiae	?	—	17	6	Azam and Kotyk, 1969; Kotyk, 1967
D-Galactose	S. cerevisiae	?	—	0.6	3	Haškovec and Kotyk, 1969; Cirillo, 1968
D-Glucose	Human red blood cells	?	—	18	8-12	Bobinski and Stein, 1969

transport properties. However, no success has yet been reported using a model phospholipid membrane and an isolated transport protein.

An interesting approach to the identification of the isolated transport proteins is the immunological one. First of all the binding of specific antibodies, produced by injecting, say, a rabbit with the purified transport protein, can help in localizing the antigen *in vivo*. This was done by Oxender's and Pardee's groups with the result that the antibodies were bound to the cell surface. This, however, does not mean that they block the transport of the substrate in question. In fact, Oxender's (with leucine transport in *E. coli*), Pardee's (with sulphate transport in *Salmonella typhimurium*) (Pardee and Watanabe, 1968) and our (with glucose transport in *Saccharomyces cerevisiae*) results indicate that the antibody is too large to have access to the binding site for the transported substrate in the intact cells (even in a yeast spheroplast), but is quite effective when applied to a relatively pure preparation of the protein (Fig. 2).

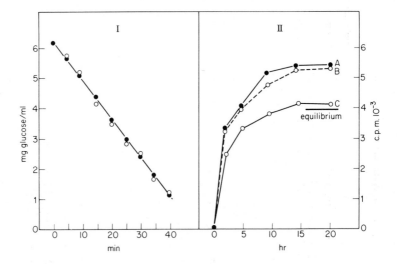

Figure 2. Effect of rabbit antiserum against glucose-carrier enriched solubilizate of baker's yeast membranes on (I) the uptake of glucose by yeast protoplasts (○, control; ●, with 1:10 antiserum added) and (II) the binding of 10^{-5} M [^{14}C] D-glucose to solubilized glucose carrier. (A) Control; (B) with 1:10 non-immune serum; (C) with 1:10 carrier antiserum.

The second approach where immunology might contribute to transport studies would be in preparing antibodies against induced cells (containing presumably the transport protein) and adsorbing them on the corresponding proteins from uninduced cells (without the transport protein), this procedure leaving only the antibody against the transport protein which might be made use of in the final purification steps.

III. IDENTITY OF THE TRANSPORT PROTEINS

Some of the isolated transport proteins have been crystallized, their molecular weight determined, and even their amino acid composition estimated (Pardee, 1968). The molecular weights of the proteins are known, most of them being about 3×10^4 daltons, and some of the molecular parameters have been defined. However, at this moment none of the transport proteins seems to represent the entire transport system for the given substrate. In all the cases where a high degree of purity of the isolated proteins has been achieved the corresponding *in vivo* transport can proceed uphill and hence, apart from the carrier itself, will contain a "permease" component through which metabolic energy for transport is provided. Moreover, the attachment of substrate to the carrier or, alternatively, its release, may be enzyme-catalysed. Genetic evidence obtained in several micro-organisms indicates the function of more than one protein in the membrane transport process (Egan and Morse 1966).

For illustrating the complexity of the transport systems in cells let me mention the work of Roseman's and Kundig's laboratory (Anderson *et al.*, 1968; Kundig *et al.*, 1964, 1966; Simoni *et al.*, 1967) where a system has been identified which provides the energy supply for the transport of some sugars in *E. coli*. This system includes a heat-stable protein, designated HPr, which can be phosphorylated by a soluble enzyme (enzyme I) forming part of the phosphotransferase system and requiring Mg^{2+} for activity. The phosphorylated P-HPr then reacts with a sugar (monosaccharide, disaccharide, as well as polyol) under catalysis of another, membrane-bound, Mg^{2+}-activated enzyme (enzyme II) which seems to be specific for each sugar and may consist of an "apoenzyme" and a specifier protein. In this reaction the sugar phosphate is translocated into the cell where it can be dephosphorylated. The role of the HPr system appears to be very significant and has been demonstrated in *Salmonella typhimurium*, *Aerobacter aerogenes* and *Staphylococcus aureus* and there are suggestions of its function in yeast cells. However, it is by no means universal since many transports, and particularly those best understood kinetically (monosaccharides in mammalian erythrocytes (Levine *et al.*, 1965; Regen and Morgan, 1964; Wilbrandt and Kotyk, 1964; Wilbrandt and Rosenberg, 1961), pentoses in baker's yeast (Cirillo, 1968; Kotyk, 1965, 1967, 1968; van Steveninck, 1966), lactose in *E. coli* (Kepes, 1960)), do not appear to involve phosphorylation of the sugar. The same holds, of course, for amino acids.

We are thus confronted with a variety of plausible transport models, most (perhaps all) involving more than one protein. In view of this complexity, it would be rather unexpected if the dissociation constants between transport protein and its substrate found *in vitro* resembled closely the half-saturation constants of transport as seems to be the case in some reports. An identity of the two constants (barring structural changes that might occur during isolation) is to be expected only if the following conditions are met.

(i) The isolated protein is the carrier whose movement in the membrane is the rate-limiting step of the transport process (this process determines the half-saturation constant of the whole transport).

(ii) It is the carrier form, such as is present at the outer face of the membrane (since the underlying concept of an active uphill transport is the change of effective affinity of the carrier as it moves from one side of the membrane to the other).

(iii) The mobility *in vivo* of the free carrier is the same as that of the loaded one (otherwise the half-saturation constant of the process is not identical with the dissociation constant of the carrier-substrate complex).

It is my conviction that it would be most useful to isolate a carrier protein from a system like that for monosaccharides in baker's yeast or in human erythrocytes, where one would not expect in the transport system the presence of any other component but the carrier and possibly the (un)coupling enzyme. However, it is rather unfortunate that these very systems show a low affinity for their substrates (generally the dissociation constant is about 10^{-3}M) which makes the isolation procedures cumbersome and less reliable than with the high-affinity systems.

It is to be hoped that in the near future, as the spectrum of isolated carrier proteins is broadened, we shall know more of the actual vital function of the membrane proteins and shall be able to reconstitute some of the transport systems in their entirety.

REFERENCES

Anderson, B., Kundig, W., Simoni, R. D. and Roseman, S. (1968). *Fedn. Proc. Fedn. Am. Socs exp. Biol.* **27**, 643.
Anraku, Y. (1967). *J. biol. Chem.* **242**, 793.
Anraku, Y. (1968). *J. biol. Chem.* **243**, 3116, 3128, 3123.
Azam, F. and Kotyk A. (1969). *FEBS Lett.* **2**, 333.
Bobinski, H. and Stein, W. D. (1966). *Nature, Lond.* **211**, 1366.
Bonsall, R. B. and Hunt, S. (1966). *Nature, Lond.* **211**, 1368.
Cirillo, V. P. (1968). *J. Bact.* **95**, 603.
Egan, J. B. and Morse, M. L. (1966). *Biochim. biophys. Acta* **112**, 63.
Fairclough, G. F., Jr. and Fruton, J. S. (1966). *Biochemistry* **5**, 673.
Fox, C. F. and Kennedy, E. P. (1965). *Proc. natn. Acad. Sci. U.S.A.* **54**, 891.
Fox, C. F., Carter, J. R. and Kennedy, E. P. (1967). *Proc. natn. Acad. Sci. U.S.A.* **57**, 698.
Haškovec, C. and Kotyk, A. (1969). *Eur. J. Biochem.* **9**, 343.
Heppel, L. A. (1967). *Science, N.Y.* **156**, 1.
Hummel, J. P. and Dreyer, W. J. (1962). *Biochim. biophys. Acta* **63**, 530.
Jones, T. H. D. and Kennedy, E. P. (1968). *Fedn Proc. Fedn Am. Socs exp. Biol.* **27**, 644.
Kepes, A. (1960). *Biochim. biophys. Acta* **40**, 70.
Kolber, A. R. and Stein, W. D. (1966). *Nature, Lond.* **209**, 691.
Kolber, A. R. and Stein, W. D. (1967). *Curr. Mod. Biol.* **1**, 244.

Kotyk, A. (1965). *Folia microbiol., Praha* **10**, 30.
Kotyk, A. (1967). *Folia microbiol., Praha* **12**, 121.
Kotyk, A. (1968). *Folia microbiol., Praha* **13**, 12.
Kundig, W., Ghosh, S. and Roseman, S. (1964). *Proc. natn. Acad. Sci. U.S.A.* **52**, 1067.
Kundig, W., Kundig, F. D., Anderson, B. and Roseman, S. (1966). *J. biol. Chem.* **241**, 3243.
Levine, M., Oxender, D. L. and Stein, W. D. (1965). *Biochim. biophys. Acta* **109**, 151.
Pardee, A. B. (1966). *J. biol. Chem.* **241**, 5886.
Pardee, A. B. (1967). *Science, N.Y.* **156**, 1627.
Pardee, A. B. (1968). *Science, N.Y.* **162**, 631.
Pardee, A. B. and Prestidge, L. S. (1966). *Proc. natn. Acad. Sci. U.S.A.* **55**, 189.
Pardee, A. B. and Watanabe, K. (1968). *J. Bact.* **96**, 1049.
Pardee, A. B., Prestidge, L. S., Whipple, M. D. and Dreyfuss, J. (1966). *J. biol. Chem.* **241**, 3962.
Penrose, W. R., Nichoalds, G. E., Piperno, J. R. and Oxender, D. L. (1968). *J. biol. Chem.* **243**, 5921.
Piperno, J. R. and Oxender, D. L. (1966). *J. biol. Chem.* **241**, 5732.
Regen, D. M. and Morgan, H. E. (1964). *Biochim. biophys. Acta* **79**, 151.
Rotman, B. and Radojkovic, J. (1964). *J. biol. Chem.* **239**, 3153.
Simoni, R. D., Leventhal, M., Kundig, F. D., Kundig, W., Anderson, B., Hartman, P. E. and Roseman, S. (1967). *Proc. natn. Acad. Sci. U.S.A.* **58**, 1963.
Stein, W. D. (1964). *In* "The Structure and Activity of Enzymes". (T. W. Goodwin, D. S. Hartley and J. I. Harris, eds.) p. 133. Academic Press, London and New York.
van Steveninck, J. (1966). *Biochim. biophys. Acta* **126**, 154.
Wilbrandt, W. and Kotyk, A. (1964). *Naunyn-Schmiedebergs Arch. exp. Path. Pharmak.* **249**, 279.
Wilbrandt, W. and Rosenberg, Th. (1961). *Pharmac. Rev.* **13**, 109.
Winkler, H. H. and Wilson, H. (1966). *J. biol. Chem.* **241**, 2200.

Reactions and Interactions in Intestinal Sugar Transport

R. K. CRANE

Rutgers Medical School, Department of Physiology,
New Brunswick, New Jersey, USA

As Dr. Kotyk has so ably reviewed during these sessions, there is currently a large effort directed toward isolation of the functional components of membrane transport systems. The effort is the more difficult because the technical approaches for carrier identification in current use seem to be pretty much limited to binding site interaction with substrates or their analogs. This restriction is largely due to the fact that the fundamental quality of transport, namely catalyzed substrate movement between two membrane-separated compartments, cannot be used for assay during isolation. As soon as isolation begins, the quality disappears because the compartments disappear. If we knew how a membrane carrier operates—whether it subserves substrate translocation through a lipoidal or an aqueous region of the membrane matrix, and whether carrier is a simple shuttle or more complex—we would stand a much better chance of isolating it, i.e. if one of these attributes could serve as the basis of an assay system or would guide us towards the selection of solvents, column components, etc. which might be useful.

Up to the present time, studies of carrier function which, unlike Na^+-K^+-dependent ATPase or bacterial sugar ester accumulation, do not involve an obvious enzyme-catalyzed reaction have been mostly limited to kinetic analysis of interaction at the two faces of the membrane. Consequently, we know a great deal about substrate interaction with carrier binding site and its influence on translocation, but we know nothing at all about the way in which translocation is accomplished.

Consequently, we have in our laboratory been very interested to find in some of our recent experiments data which suggest that we have seen a little way beyond substrate-binding site interaction into the mystery of translocation. We have accumulated data which allows us the interpretation that there is an intermediate step in the translocation of simple sugars by the Na^+-dependent carrier system of the intestines' brush border membranes; a step in addition to, or beyond, at least in time, the formation of enzyme-carrier complex at the external side of the membrane.

As our analysis of the data is based upon simple mobile carrier theory, it is just as well to remind you what that theory is and what it implies. The system we are studying, i.e. intestinal sugar transport, operates kinetically as depicted in Fig. 1. Substrate and Na^+ undergo Michaelis-Menton association with individual binding sites on a single membrane component to form a ternary complex. The complex reorients so that Michaelis-Menton dissociation can occur into the other compartment. This carrier is totally equilibrational, i.e. movement of substrate up a gradient into the cells requires the simultaneous movement of Na^+ down a gradient maintained by continuous ejection of Na^+ at another cell locus. Our experiments have suggested that carrier sites reorient freely across the membrane whether one or both substrates are present or absent.

This formulation is obviously in agreement with the simplest of carrier theory. What is not so obvious is the corollary to this formulation, and to simple carrier theory as well, namely that interaction of substrate with the carrier must

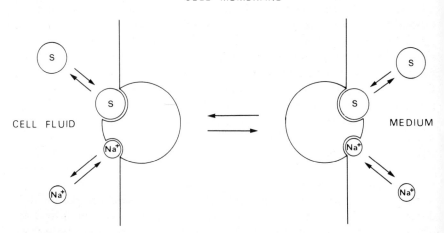

Figure 1. Diagrammatic representation of the operation of a Na^+-dependent bifunctional mobile carrier. (From Crane, 1965.)

result in translocation of substrate; unless the particular substrate used forms an abnormal complex or has alternative interactions with other parts of the membrane which may prevent carrier reorientation or movement.

We have now found an instance, i.e. a particular compound for which translocation does not follow upon an interaction which appears by every available measure to be normal and for which alternative interactions are highly improbable. We interpret these findings as evidence for a second step in transport.

The compound we have studied is 6-deoxy-L-galactose (L-fucose) which we had earlier noticed with some surprise to be a good inhibitor of sugar transport (Crane, 1960). It was surprising because the structure of 6-DLGal does not compare with that of a good substrate, e.g. glucose, but compares more with that of mannose, which is an exceedingly poor substrate and is not, itself, an inhibitor of any note. We did not study 6-DLGal any further at the time but turned our efforts to understanding the nature of the Na^+-ion dependency of the transport. As I believe we now do, at least in major respects, we turned again to 6-DLGal.

First, we studied the characteristics of entry of 6-DLGal and found that it did not enter the tissue, at least to any appreciable extent. Under conditions such that good accumulation of substrates like glucose would have been seen, 6-DLGal tissue space barely exceeded that of the extracellular marker, mannitol, and was considerably less than that of 2-deoxyglucose which does not interact with the glucose system. Phlorizin, a potent competitive inhibitor of glucologue transport had, of course, no effect on the mannitol space nor the entry of 2-DG but it also had no effect on the distribution of 6-DLGal. Hence, whatever small degree of entry of 6-DLGal there may be, it is probably not by way of the glucologue carrier.

We next investigated the nature of 6-DLGal inhibition of glucologue transport. Based on carrier theory as we understand it, we were prepared to see 6-DLGal emerge as a non-competitive inhibitor. Experiment proved otherwise. Using L-glucose, 3-O-methyl-D-glucose or 1,5-anhydro-D-glucitol (1-deoxyglucose) as substrates, 6-DLGal proved to be a competitive inhibitor with a K_i value of about 20mM. It interacts with the carrier binding site about three times better than L-glucose, about equally with 1-DG and somewhat less well than 3-MG.

In addition to competitive inhibition, another test for common substrates of a mobile carrier system is the induction of counterflow of one substrate by another. Our success, for example, in inducing counterflow of L-glucose by the addition of D-glucose was an important aspect of the proof that L-glucose is a substrate for the intestinal glucologue system (Caspary and Crane, 1968). In the present situation, we have a competitive inhibitor which does not enter the tissue. Hence, we have looked not for 6-DLGal to induce counterflow but rather for its failure to do so. Carrier theory requires the entry of the inducer. We committed a substantial effort to this question, particularly owing to the fact that it was a negative result we looked for and it was a negative result we always found. In the most convincing of the experiments, 6-DLGal was compared with 1-DG as elicitor for the counterflow of L-glucose. The one appears to be a non-penetrating competitive inhibitor, the other is an entering substrate; as inhibitors they have about the same K_i. In this experiment, 1-DG elicited counterflow of L-glucose; 6-DLGal was devoid of measurable effect.

Under ordinary circumstances, with a simple mobile carrier, the needed proof would be substantially complete at this point. However, we are not working with a simple, monofunctional mobile carrier such as the glucose carrier of the red cell membrane. We are working with a bifunctional mobile carrier (Fig. 1) which has two specific sites which interact with one another. Binding of Na^+ increases the substrate affinity of the glucose site (Crane, 1965). Binding of glucose increases the substrate affinity of the Na^+ site (Lyon and Crane, 1966). Consequently, the possibility had to be considered that 6-DLGal interacted with the substrate site in an abnormal manner, i.e. in such a way that interaction of Na^+ with its site was prevented or modified. If 6-DLGal formed a complex abnormal with respect to Na^+ binding, then the failure of its transport could be explained in that way. However, 6-DLGal does not form an abnormal complex. To achieve this answer we turned to measurements of transmural potential, one of the most sensitive methods for detecting alterations in Na^+ flux. It is well known that addition of substrates of the glucologue carrier to the mucosal side of an *in vitro* preparation of intestine elicits an increase over the resting transmural potential which is positive to the serosal side. Inasmuch as the increase has been shown to be matched by a change in short circuit current and the directly measured flux of Na^+, this phenomenon has been understood in terms of, and as substantiating evidence for, the model illustrated in Fig. 1. It is also well known that under the same *in vitro* conditions addition of a non-substrate for the Na^+-dependent carrier will, on the contrary, reduce the transmural potential in proportion to the concentration of the compound added. This effect is interpreted as an osmotic streaming potential of the opposite sign to the transmural potential due to the flow of water through the charged mucosal membrane. Some typical results are shown in Table 1.

It may be seen that the substrates, D-glucose, 1-DG, 3-MG, and L-glucose all elicit an increase in the transmural potential and by concentrations in approximate proportion to their relative K_m values. Mannitol, on the other hand, elicits a negative potential and 6-DLGal exactly matches this effect. 6-DLGal binds but it neither enters nor induces an entrance of Na^+.

We are ready then to make the test of whether 6-DLGal forms a normal complex. Since the binding of Na^+ increases the affinity of the substrate site, the K_m value is dependent upon the Na^+ concentration in the suspending medium. It follows then that if 6-DLGal forms a normal complex, its K_i value will be dependent upon the Na^+ concentration. More than that, if a normal complex is formed, Na^+ concentration should have quantitatively the same effect on 6-DLGal as it has on substrate. Experimentally, what should then be found is that the K_m of a substrate and the K_i of 6-DLGal change with changing Na^+ concentration, but that the ratio, K_m/K_i, remains constant. As shown by the data in Table 2, a constant K_m/K_i ratio was, indeed, what we found. So far as we can tell, 6-DLGal forms a completely normal complex with the carrier.

With this information in hand, we are then in a position to consider its meaning. 6-DLGal forms a normal complex but it is not transported; the complex is abortive. Inspection of Fig. 1 or of any of the extant models of

Table 1. Effect of various sugars and D-mannitol on the transmural PD in hamster small intestine

Substrate (mM)	Substrate induced potential difference (PD) (mV)
D-glucose (4)	
1	+ 5.5
5	+ 9.6
10	+ 10.1
1,5-anhydro-glucitol (3)	
1	+ 1.8
5	+ 4.8
10	+ 7.2
3-0-methyl-glucose (4)	
10	+ 5.6*
20	+ 7.0*
30	+ 8.1*
L-glucose (3)	
5	+ 0.3*
10	+ 0.5*
20	+ 0.9*
6-Deoxy-L-galactose (3)	
10	− 1.2†
20	− 2.1†
30	− 3.8†
D-mannitol (3)	
10	− 1.0
20	− 2.2
30	− 3.5

PD values were obtained 30 sec after transfer and are averages of three or four experiments. The number of tests is given in parentheses.
* Corrected for D-mannitol osmotic streaming potential.
† Not corrected for D-mannitol osmotic streaming potential.

carrier transport in similar form does not tell us why. There is no provision made in these models for anything beyond Michaelis-Menton interaction except translocation, and it follows from our observations that translocation must be more complex than the terms "binding site reorientation" or "movement" imply. There must be a step in the process which the 6-DLGal-Na$^+$-carrier complex cannot undergo. Alternative interactions of 6-DLGal are most unlikely—they are not seen in many other compounds similar in shape, size and hydrogen-bonding capacity. To accentuate the point, alternative interactions are seen with phlorizine, i.e. the phloretin moiety binds to the membrane, and phlorizin acts like 6-DLGal. It is a competitive, non penetrating substrate but it is

Table 2. Influence of different medium-Na^+ concentrations on K_m of 3-O-methylglucose and K_i of 6-deoxy-L-galactose determined by measurements of transmural PD* in hamster small intestine

Na^+ concentration (mequiv./L)	$Tris^+$ concentration (mequiv./L)	K_m of 3-O-methyl-glucose (M x 10^{-3})	K_i of 6-DLGal (M x 10^{-3})	Ratio K_m/K_i†
145	–	22	8.5	2.6
85	60	25	10.5	2.4
40	105	39	17	2.3
31	114	66	24	2.7

* PD values were recorded in Krebs-Henseleit bicarbonate buffer 30 sec after transfer.
† Mean ± standard deviation = 2.50 ± 0.22.

understandable as such because of the alternative interaction. 6-DLGal has none, or none that can be detected.

We also, over the years, have been unable to find evidence of any bonding interactions, other than hydrogen bonds, during sugar transport in the intestine. Hence, we cannot say that 6-DLGal is unable to undergo a reaction of transport. However, we can say that the 6-DLGal-carrier complex is unable to undergo a rearrangement—a conformational alteration—which must precede or occur during translocation. It is substantially the only deduction that the data of many years leaves available to us. It is this deduction which is illustrated in Fig. 2.

It is well known that conformational rearrangements such as the mechanism of transport have been postulated a number of times and in a variety of guises. They are all reasonable postulates. To my knowledge, however, this is the first time the suggestion has been made according to the demands of the data. Even so, it is not a particularly new suggestion for us. This postulate for sugar transport clearly resembles our postulate of some 8-9 years ago; namely, that in the hexokinase reaction, a conformational rearrangement of the substrate-enzyme complex follows Michaelis-Menton interaction and precedes phosphoryl transfer (Crane, 1962).

To get back to the beginning, the question is how can such information help us or guide us toward isolation. Let us assume that there is, indeed, a conformational rearrangement, then let us ask what its purpose might be. It might, of course, be that step of translocation, of reorientation, itself. If it is, we are not helped very much. However, it may not be that, but rather a step preceding translocation, and if it is help may be forthcoming.

There is now available a well-developed, immensely reasonable concept of passive diffusion of "lipid-soluble" compounds through membranes as governed by their degree of interaction—hydrogen-bond interaction—with water and the necessity that these bonds be broken for transfer to occur (Stein, 1967). It is only a small extension of this view to say that sugars require carriers because their interactions with water are so numerous, and a still smaller extension to imagine that carriers are able to act because they replace the interactions (H-bonds) of sugars with water by interactions (H-bonds) with themselves. By this line of thinking one is led quite naturally to recall the now well-documented

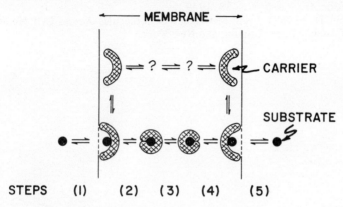

Figure 2. Diagrammatic representation of the steps in sugar translocation. Na^+ has been omitted for clarity. (From Caspary et al., 1969.)

action of ionophores in providing associations with cations in replacement of the normal association of cations with the water of their environment with the formation of a lipophilic ion-carrier complex (Pressman et al., 1967).

Once one has reached this point the conclusion becomes all too obvious. Is it not credible that the sugar-carrier complex does, indeed, undergo a conformational rearrangement in order to create, like the ionophores with their cations, a membrane-soluble form of sugar which can easily cross the lipoidal matrix and release sugar to the aqueous phase on the other side by a reversal of the process? Should we, in our search for carrier, look more intensively for a relatively small molecular weight protein, enriched in lipophilic side chains which has the detectable property of increasing the lipid-solubility of certain but not all sugars (glucose, say, and not 2-deoxyglucose) if we are searching for the glucologue carrier of the intestine? I am, perhaps, too much persuaded by the simplicity of biology which would be indicated by such a conclusion. Nonetheless, now that we can see something that looks like a second step, and we can imagine an adequate reason for such a second step, it would seem a shame not to try it out.

REFERENCES

Caspary, W. F. and Crane, R. K. (1968). *Biochim. biophys. Acta* **163**, 395.
Caspary, W. F., Stevenson, N. R. and Crane, R. K. (1969). *Biochim. biophys. Acta* **193**, 168.

Crane, R. K. (1960). *Physiol. Rev.* **40**, 789.
Crane, R. K. (1962). *In* "The Enzymes" (edited by P. D. Boyer, H. Lardy and K. Myrbäck), Vol 6, p. 47. Academic Press, London and New York.
Crane, R. K. (1965). *Fedn Proc. Fedn Am. Socs exp. Biol.* **24**, 1000.
Lyon, I. and Crane, R, K. (1966). *Biochim. biophys. Acta* **112**, 278.
Pressman, B. C., Harris, E. J., Jagger, W. S. and Johnson, J. H. (1967). *Proc. Natn. Acad. Sci., U.S.* **58**, 1949.
Stein, W. D. (1967). "The Movement of Molecules Across Cell Membranes". Academic Press, London and New York.

Sucrase and Sugar Transport in the Intestine: A Carrier-like Sugar Binding Site in the Isolated Sucrase-Isomaltase Complex

G. SEMENZA[†]

*Biochemisches Institut der Universität Zurich, Switzerland,
and Istituto di Fisiologia Generale dell'Università di Milano, Italy*

Doctor Kotyk has summarized and discussed the approaches which are being used in the purification and identification of membrane proteins involved in transport (Kotyk, 1970). Successful as they are when applied to microbial systems, they bear little hope with systems from higher organisms as suitable mutants are not easily available and most membrane transport systems have large apparent "transport K_m and K_i values", etc.

Our own approach has developed along the following lines: (a) sucrase and sugar carrier are located close to one another in the brush borders of small intestine; (b) they are affected by monovalent cations in much the same way and may even share the same Na^+-binding site(s); and (c) we should therefore investigate the possibility of whether the artificially solubilized and isolated membrane fragment which contains sucrase also contains the substrate-binding site of the sugar carrier. We shall discuss these points in sequence.

I. THE PROXIMITY OF SUCRASE AND SUGAR TRANSPORT SYSTEM IN THE SMALL INTESTINE

In 1961 Miller and Crane investigated the uptake of glucose by hamster small intestine using *sucrose* as a substrate. Higher tissue-glucose/medium-glucose ratios were obtained than with free *glucose* as a substrate; moreover, addition of glucose-oxidase to the medium did not reduce the uptake of glucose when sucrose was the substrate.[*] This kinetic advantage of the glucose moiety of sucrose over free glucose can be explained, in principle, in a number of ways.

It is possible, for example, that the anomeric form of glucose liberated by sucrase is absorbed by the small intestine better than the anomeric form of glucose prevailing at the equilibrium. But this is not so because sucrase liberates

---

[*] Their observations were confirmed and extended to other species and for maltose by other authors (e.g. Rutloff *et al.*, 1965; Ugolev, 1968; Parsons and Prichard, 1968).

[†] Present address: Laboratorium für Biochemie der E.T.H., Universitätstrasse 6, 8006 Zurich, Switzerland.

glucose as alpha (Fig. 1) (Semenza *et al.*, 1967), whereas both forms of glucose are absorbed equally well (Table 1) (Semenza, 1969a).

Another possibility which may be excluded is that sucrase itself may act as a carrier for glucose. In fact, tris-hydroxylmethyl-amino-methane (Tris) competitively inhibits sucrase (Kolínská and Semenza, 1967) but does not affect sugar

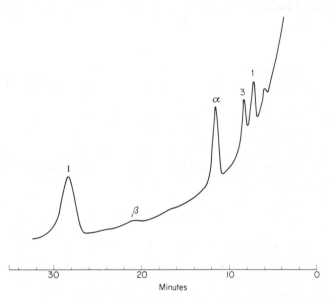

Figure 1. Gas-liquid chromatogram of the monosaccharides (as persilylated derivatives) liberated by intestinal sucrase from sucrose after 2 min incubation. α, α-glucopyranose; β, β-glucopyranose; 1 and 3, fructose derivatives; I, internal standard (inositol). From Semenza *et al.* (1967).

uptake (Bosačková and Crane, 1965). Conversely, phlorizin, which competitively inhibits sugar uptake, does not affect sucrase (Semenza, 1969b). Moreover, animals devoid of small-intestinal sucrase activity (e.g. most mammals before weaning (for review, see Semenza, 1968), frogs (Parsons and Prichard, 1965), humans suffering from sucrose-isomaltose malabsorption (Anderson *et al.*, 1963; Auricchio *et al.*, 1965; Semenza *et al.*, 1965)) do absorb monosaccharides. Finally, the apparent K_m value for glucose uptake in the small intestine does not change during the development of intestinal sucrase in baby rats (Table 2) (Colombo *et al.*, 1969).

At least two possibilities remain: (a) that sucrase acts as a transglucosidase by transferring the glucosyl moiety of sucrose onto the carrier; and (b) that a local hyperconcentration of glucose builds up in the immediate neighbourhood of the carrier during the hydrolysis of sucrose. The latter explanation is actually the one Miller and Crane (1961) originally suggested. However, either possibility

Table 1. Uptake of sugars in hamster small intestine from the mucosal side at 37°C (from Semenza, 1969)*

		Starting from			
Sugars	n	α form	Equilibrium mixture	β form	P†
6-Deoxy-D-glucose	25	2.74 ± 0.77	3.04 ± 0.62		< 0.02
3-O-Methyl-D-Glucose	15	1.81 ± 0.54	1.71 ± 0.69		n.s.
D-Glucose (5 mM)	9	2.56 ± 0.86	2.57 ± 0.72		n.s.
D-Glucose (at 25°C)	9	1.53 ± 0.93	1.73 ± 1.04		n.s.
D-Glucose (5 mM)	15	5.27 ± 1.35		5.76 ± 1.38	n.s.

Uptake is given in μmoles/min^{-1}/ml^{-1} tissue water ± S.D. Unless stated otherwise, the substrates were 10 mM.
* Doctor Faust draws my attention to a former paper of his where he described very similar results (Faust, 1964).
† Calculated from the differences in uptake in paired experiments (n.s. = not significant).

Table 2. Sucrase activity and apparent Michaelis constants for glucose uptake in the small intestine of rats of different ages

Age (days from birth)	n	Apparent Michaelis constant for glucose uptake (mM ± standard deviation)	Sucrase activity (Units/g fresh weight)
43.5 (22-68)	12	6.1 ± 3.1	1.46
15.5 (14-18)	3	4.1 ± 1.0	0.29

implies that sucrase and sugar carrier are located close to each other—how "close", however, we still do not know.*

* Hamilton and McMichael (1968) and Prichard (1969) have suggested that the "fuzz" is responsible for trapping the liberated glucose in the inter-microvillal space. Although this mechanism may certainly play a role, it seems unlikely to be the only one, because the glucose residue of glucose-6-phosphate does not show any kinetic advantage over free glucose (Miller and Crane, 1961). The enzyme splitting glucose-6-phosphate (alkaline phosphatase) (Ito, 1969) and sucrase (see below) are both localized in the membrane of the brush borders.

II. Na^+ ACTIVATION OF SMALL INTESTINAL SUCRASE AND OF SUGAR UPTAKE

Not only intestinal sugar uptake (Crane, 1962), but also sucrase, is activated by Na^+ (Semenza et al., 1963, 1964; Semenza, 1967a,b, 1968; 1969b).*

A comparison of the effects of monovalent cations on sucrase and sugar uptake in some species has shown a close parallelism. This parallelism indicates either that the Na^+ site(s) of sucrase and the sugar carrier have similar chemical properties, or that the same Na^+ site(s) are shared by sucrase and the sugar carrier. These observations were discussed in previous publications in some detail (Semenza, 1967a,b, 1968, 1969b).

The solubilization and isolation of intestinal sucrase has been reported by Kolínská and Semenza (1967). Because of the close proximity of sucrase to the sugar carrier and the possibility of common Na^+-binding site(s) we have investigated whether the solubilized sucrase-isomaltase complex also contains carrier-like substrate binding site(s).

III. A PRELIMINARY EXPERIMENT

If papain solubilizes the Na^+-dependent sugar carrier (or a part of it) along with sucrase, it must correspondingly reduce the Na^+-dependent sugar uptake in addition to increasing the so-called "extra-cellular" space.

Figure 2 shows that this is actually the case. Two procedures for sucrase solubilization were tested: one is very effective (papain digestion, at left), the other is comparatively ineffective in the hamster (Triton X-100, at right). The data on 3-methyl-glucose uptake are corrected for the increased 2-deoxy-glucose space (reported at the bottom). It is apparent that neither agent can solubilize sucrase without reducing Na^+-dependent sugar uptake. With papain, sucrase solubilization and disappearance of Na^+-dependent sugar uptake run almost parallel. Although the results are compatible with the mechanism suggested at the beginning of the paragraph other explanations may also be possible.

IV. GLUCOSE FIXATION BY THE ISOLATED SUCRASE-ISOMALTASE COMPLEX

The fragment of brush border's membrane containing the sucrase-isomaltase complex was solubilized and isolated according to an extensive modification of a procedure which was published previously (Kolínská and Semenza, 1967). Briefly, sucrase was solubilized by papain digestion and isolated by Sephadex chromatography. The protein eventually obtained fulfilled the standard criteria

---

* In the following, unless stated otherwise, the Na^+-dependent trans-membrane sugar transport system of the small intestine is referred to as "the sugar transport system".

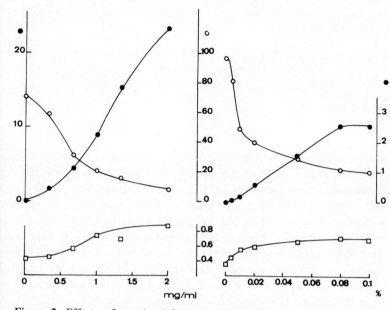

Figure 2. Effects of papain (left) and of Triton X-100 (right) on hamster small intestine. Small-intestinal rings were incubated for 15 min at 37°C in Krebs-Henseleit buffer in the presence of papain + cysteine. HCl (left, concentration of papain on the abscissae), or in the presence of Triton X-100 (right, concentration of Triton on the abscissae). The rings were then washed and incubated again for 15 min at 37°C in Krebs-Henseleit buffer, containing 2 mM 3-O-[^{14}C]methyl-D-glucose + 1.5 mM 2-deoxy-D-glucose. After the second incubation, the rings were homogenized and deproteinized with $ZnSO_4$ + $Ba(OH)_2$. The following parameters were determined: the units of sucrase solubilized during the first incubation (●); the "2-deoxy-D-glucose space" during the second incubation (□); the uptake of 3-methyl-glucose during the second incubation (○); the latter data are corrected for the 2-deoxy-glucose space.

Table 3. Composition of the incubation medium for glucose-binding experiments

Tris. Cl	80 mM
NaCl	123 mM
K phosphate buffer, pH 7.4	3.7 mM
Glucose	0.2 – 2 mM

of purity, such as one single band in disc- and starch gel-electrophoresis, and in ultracentrifugation procedures.

We were mainly aware of four possible sources of error:

(a) sucrase and isomaltase, being glucohydrolases and glucosyltransferases, were expected to bind glucose at the substrate binding site(s). Therefore, glucose binding was measured in the presence of high concentrations of tris-hydroxymethyl-amino-methane, a fully competitive inhibitor of these carbohydrases (Kolínská and Semenza, 1967) (Table 3). Increasing the

Tris concentration further did not result in any decrease of glucose fixation (Table 4).

Table 4. Lack of effect of Tris on the binding of glucose onto sucrase-isomaltase ([Glucose]: 0.2 mM)

Controls (in 80 mM Tris)	100%
in 160 mM Tris	84 ± 19%
in 240 mM Tris	97 ± 29%

(b) sucrase-isomaltase, as well as other proteins tested, can bind glucose "non-specifically". This source of error was reduced, but probably not eliminated, by working at a rather low glucose concentration —0.2 to 5 mM— in the hope that the affinity of the looked-for "carrier-like glucose-binding site" (CLGBS) for glucose would be higher than that of other, "non-specific" binding sites. Nevertheless, fixation at sites other than the CLGBS made it impossible to estimate both the maximum number of glucose molecules fixed at the CLGBS and its apparent "binding-K_m". It should be noted, however, that it is still unclear how large the apparent "transport-K_m" for glucose actually is (*in vitro* vs. *in vivo* experiments, see Parsons and Prichard, 1966). Furthermore, "transport K_m values" can be expected to match "binding K_m values" only if a number of conditions are fulfilled (see, for example, Kotyk, 1970).

(c) Bacterial contamination occurring either before or after the beginning of the incubation was ruled out by filtering the sucrase-isomaltase complex through a Millipore filter prior to incubation and by adding antibiotics to the incubation medium. Neither procedure affected glucose binding.

(d) Dialysis tubing and all filters made of cellulose derivatives bind glucose reversibly. Although the sucrase-isomaltase preparations fulfilled the standard criteria of purity, one could not rule out *a priori* the presence of a very minor impurity having large glucose-binding capacity, particularly since both dialysis and Sephadex chromatography were used in purifying sucrase. The effect of pretreatment with trypsin on glucose fixation was therefore tested (Table 5). It is apparent that glucose binds to a proteinaceous component; the data do not convey any information, however, on whether glucose actually binds to amino acid residues or, for example, to a carbohydrate moiety.

The following characteristics of glucose binding onto the sucrase-isomaltase complex were investigated:

(a) *Reversibility.* Bound radioactivity is lost upon dilution with water or with a solution of cold glucose.

(b) *Na^+ dependence.* Substituting K^+ for Na^+ decreases the glucose-binding capacity (Table 6).

Table 5. Effect of urea + trypsin pretreatment on glucose binding onto sucrase-isomaltase

	nmoles of glucose bound	%
No pretreatment	3.67	100
Pretreatment with urea alone	3.66	99.6
Pretreatment with urea + trypsin	0.04	7

Sucrase-isomaltase was first incubated for 20 min at room temperature in the absence or presence of 5M urea; 1.25 mg of trypsin were added to a volume of 25 μl, and the incubation pursued for a further 30 min; the samples were diluted 1:5 with water, and after 30 min the glucose binding capacity was tested ([glucose]: 0.27 mM).

Table 6. Na^+ dependence of $[^{14}C]$ glucose binding onto the sucrase-isomaltase complex from rabbit (glucose concentration: 0.27 mM; nmoles of glucose bound per nmole of sucrase (average of six experiments ± S.D.))

In 123 mM Na^+, 7 mM K^+	3.42 ± 0.81
In 0 Na^+, 130 mM K^+	0.77 ± 0.54

Similar results were obtained with the sucrase-isomaltase complex from hamster.

(c) *Inhibition or lack of inhibition by sugars and sugar derivatives* (Table 7 and Fig. 3). The effect of some representative sugar and sugar derivatives was studied at low glucose concentrations (0.27 to 2.07 mM) for the reasons reported above, and at high "inhibitor" concentrations, due to the uncertainty of the real "K_m values". It is apparent that most sugars and sugar derivatives which are transported by the Na^+-dependent system (6-deoxy-D-glucose, 3-deoxy-D-glucose, 3-methyl-D-glucose, α-methyl-D-glucopyranoside, D-fucose, arbutin) lower glucose fixation onto sucrase-isomaltase; sugars which are absorbed poorly or not at all by this system have little or no effect (D-xylose, 2-deoxy-D-glucose, D-mannose, D-fructose).

Besides minor quantitative differences, the major discrepancies between transport systems and glucose-binding are shown by D-galactose and phlorizin, neither of which apparently affects glucose binding. A number of explanations are possible for this. Newey *et al.* (1965) have suggested that glucose is absorbed in the intestine by two transport systems, only one of which is shared with galactose. Furthermore, Alvarado (1966, 1968) has reported that galactose, but not glucose, inhibits amino acid uptake allosterically by binding at a site identical with that of its own transport (in hamster, small intestine). It is thus possible that glucose and galactose bind to two distinct (possibly overlapping) sites. Sucrase-isomaltase complex would contain the glucose-binding site only.

Table 7. Effect of various sugars and sugar derivatives (100 mM) on glucose fixation by rabbit sucrase-isomaltase (the results are expressed as per cent of the paired controls ± S.D. (in brackets the number of experiments))

	Glucose concentration		
	0.27 mM	0.47 mM	2.07 mM
Controls	100%	100%	100%
6-Deoxy-D-glucose (5)		44 ± 21	
3-Deoxy-D-glucose (4)		0	0
3-O-Methyl-D-glucose (5)			22.5*
α-Methyl-D-glucopyranoside (5)			16.8 ± 11.7
6-Deoxy-D-galactose (4)		0	0
Arbutin (2)			20.5
D-Xylose (4)		54	no inhibition
6-Deoxy-L-galactose (4)	0		
D-Galactose (7)		112 ± 17.5	119 ± 8
2-Deoxy-D-glucose (2)			71.3†
D-Mannose (9)		no inhibition	107 ± 13
D-Fructose (4)		134 ± 17	
m-Inositol (4)	73‡		
Phlorizin, 0.1 mM (8)	120		118

* Complete inhibition at 200 mM (3).
† 34.1 (2) at 200 mM.
‡ 83 (4) at 50 mM.

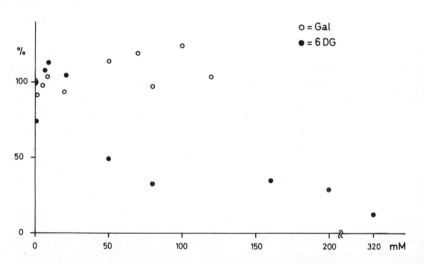

Figure 3. Effect of increasing concentrations of 6-deoxy-D-glucose (●) or of D-galactose (○) (concentrations on the abscissae) on glucose binding onto sucrase-isomaltase. The concentration of glucose was 0.54 mM in the experiment with 6-deoxy-glucose, and 0.47 mM in the experiment with galactose.

Alternatively, the discrepancy between the substrate specificity and the effect of glucose binding onto the sucrase-isomaltase complex may be ascribed to an alteration arising during solubilization, and/or to the complexity of the transport system, which need not be a one-step system (see Kotyk, 1970). However, most if not all the membrane proteins isolated from micro-organisms, and supposed to take part in membrane transport, also show discrepancies similar to the one which we observe for sucrase-isomaltase.

With phlorizin, the mechanism whereby this glucoside inhibits intestinal uptake with such a small K_i (approx. 10^{-6} M) is still obscure. Two hypotheses have been put forward: according to one (Diedrich, 1966; Alvarado, 1967) phlorizin binds to *two* membrane sites, one of which is the sugar carrier. According to the other (Diedrich, 1968), the actual inhibitor is not phlorizin, but its aglycone (phloretin), once it is liberated in the immediate neighbourhood of the sugar carrier. Either hypothesis provides a logical explanation as to why phlorizin does not prevent glucose binding onto sucrase-isomaltase: the two phlorizin-binding sites may have been separated during solubilization; on the other hand, isolated sucrase-isomaltase does not have any phlorizin-hydrolase activity.

We found a clear correlation between the extent of inhibition of 6-deoxy-D-glucose uptake by phlorizin and the amount of phlorizin-hydrolase activity in the intestine (Fig. 4) (Colombo *et al.*, 1969). These data, however, do not necessarily imply that phloretin, as suggested, is the actual inhibitor (this point will be discussed by Alvarado in the next paper), but merely indicate that a "phlorizin-hydrolase" activity is necessary for phlorizin inhibition. The clarification of the mechanism of phlorizin action may come through a better knowledge of the properties of "phlorizin-hydrolase". In the meantime we should not take offence if this glucoside has no effect on glucose binding by the sucrase-isomaltase complex.

(d) *Effect of anti-sucrase antibodies.* Antibodies were prepared against the sucrase-isomaltase complex (Cummins *et al.*, 1968). They do not inhibit sucrase activity. However, these immune sera and the IgG prepared therefrom (according to Fahey, 1967 and Deutsch, 1967) inhibit intestinal uptake of 6-deoxy-D-glucose *in vitro* (Table 8). This inhibition is clearly not the result of cytolysis, because it is even more evident after inactivation of the complement and because the "extracellular" space is unaffected by these additions.

Steric hindrance (due, for example, to the proximity of sucrase and sugar carrier) is unlikely to affect low mol. wt. substrates in immunological systems (Cinader, 1967). The inhibition of sugar uptake by anti-sucrase antibodies seems to indicate, rather, that the sucrase-isomaltase complex contains a component antigenically related to a part of the sugar transport system. It further indicates that sucrase itself (the activity of which is unaffected by these antibodies) is not the carrier (see above).

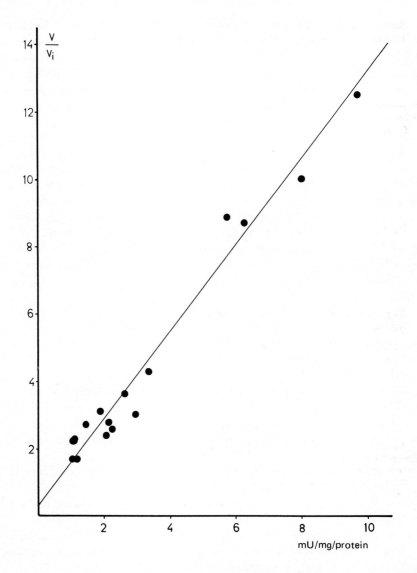

Figure 4. Correlation between inhibition of 6-deoxy-glucose uptake in rat small intestine by phlorizin (ordinate) and intestinal phlorizin-hydrolase activity (abscissae). v, 6-deoxy-glucose uptake in the absence of phlorizin; v_i, 6-deoxy-glucose uptake in the presence of 0.01 mM phlorizin mU, milliunits of phlorizin-hydrolase activity, as determined with the method of Malathai and Crane (1969). From Colombo *et al.* (1969). The line is the calculated regression line.

Table 8. Inhibition of 6-deoxy-D-glucose uptake in rabbit small intestine by antibodies directed against rabbit intestinal sucrase-isomaltase complex.

	6-Deoxy-glucose uptake	Difference from the controls ± S.D.	P*	Extra-cellular space
Total immune serum (4)				
Controls	1.15 (100%)			0.14
⁺immune serum	1.0 (88%)	12 ± 9.3	n.s.	0.14
⁺pre-heated immune serum (56°C for 30 min)	0.92 (76%)	24 ± 13	0.05	0.14
Total non-immune serum (4)				
Controls	1.13 (100%)			0.11
⁺serum	0.99 (87%)	13 ± 16.5	n.s.	0.12
⁺pre-heated serum (as above)	1.14 (98%)	2 ± 19	n.s.	0.11
IgG (7)				
Controls	2.38 (100%)			0.10
With IgG	2.16 (92%)	8 ± 12.6	n.s.	0.10
With pre-heated IgG (as above)	1.97 (82%)	18 ± 9.69	0.01-0.001	0.10
Non-immune gamma-globulin (6)				
Controls	1.47 (100%)			0.09
With pre-heated non-immune gamma-globulin (as above)	1.38 (94%)	6 ± 21	n.s.	0.09

Incubation at 37°C for 10 min in Krebs-Henseleit buffer. Intestine mounted in frames. The amount of immune serum added was calculated to react with most of the sucrase present in the incubation mixture. The data are corrected for the extracellular space (2-deoxy-glucose space) and expressed as μmoles/10 min^{-1}/ml^{-1} tissue water (in brackets the number of experiments).

* Calculated from the difference from the controls in paired experiments (n.s. = not significant).

Does glucose-binding on to sucrase-isomaltase complex indicate the presence of carrier's substrate-binding site in it? As discussed in the previous paragraphs, the data are compatible with this idea, although the substrate-binding site may have been partially modified during solubilization. Glucose binding on to sucrase-isomaltase fulfills a number of criteria and conditions which one would expect to be fulfilled by the carrier's substrate-binding site:

(a) sucrase-isomaltase is certainly a membrane component (for a review, see Semenza, 1968) and proteolytic digestion is necessary to bring it into solution (Fig. 2). Recent e.m. studies with negative staining (Nishi *et al.*, 1968) and with ferritin-labelled antibodies (Gitzelmann *et al.*, 1970) also agree with this conclusion. It should be noted that this apparently trivial condition is not fulfilled by at least one glucose-binding component in the brush borders (Faust *et al.*, 1968) which was shown unequivocally by Eichholz (1969) to occur in the *cores* and not in the membranes of the brush borders.

(b) During solubilization of sucrase-isomaltase there is a parallel decrease in the ability of the intestine to absorb sugars actively (Fig. 2).

(c) This glucose-binding component occurs close to sucrase, as expected (see above).

(d) Glucose binding is reversible.

(e) It is Na^+ dependent. If this Na^+ dependent glucose-binding is indeed due to the carrier, our observations would provide direct evidence favouring Crane's hypothesis (1962) vs. Csáky's hypothesis (1963) on the mechanism of Na^+ activation of sugar intestinal uptake.

(f) It is inhibited by most sugars which show Na^+ dependent absorption and it is unaffected by other sugars.

(g) Antibodies directed against the sucrase-isomaltase complex (which do not inhibit sucrase activity) inhibit sugar uptake *in vitro*.

Do these considerations suffice to answer the title question in a positive sense?

All above-mentioned criteria are indirect in nature. In the absence of functional evidence it is impossible to exclude the possibility that the sucrase-isomaltase complex may contain a structure similar to the substrate-binding site of the sugar carrier but not itself be involved in sugar transport. This possibility is very real because many (presumably phylogenetically-related) proteins are known which have identical partial structures, e.g. trypsin and chymotrypsin, lactalbumin and lysozyme, alpha and beta chains of hemoglobin. It should be noted that this severe limitation applies equally well to many of the "transport proteins" isolated from micro-organisms.

ACKNOWLEDGEMENTS

The present work was carried out in collaboration with A.-K. Balthazar, V. Colombo, R. Gitzelmann, T. Koide, J. Kolínská, J. Lindenmann, H. Mosimann and E. Mülhaupt. The financial support of the Schweiz. Nationalfonds zur Förderung der wissenschaftlichen Forschung, Bern; of the Schweiz. Akademie der mediz. Wissenschaften; of the Air Force Office of Scientific Research through the European Office of Aerospace Research O. A. R., United States Air Force (Grants AF EOAR 67-17 and EOOAR-68-0017), is gratefully acknowledged.

REFERENCES

Alvarado, F. (1966). *Science, N.Y.* **151**, 1010.
Alvarado, F. (1967). *Biochim. biophys, Acta* **135**, 483.
Alvarado, F. (1968). *Nature, Lond.* **219**, 276.
Anderson, C. M., Messer, M., Townley, R. R. W. and Freeman, M. (1963). *Pediatrics*, **31**, 1003.
Auricchio, S., Rubino, A., Prader, A., Rey, J., Jos, J., Frézal, J. and Davidson, M. (1965). *J. Pediat.* **66**, 555.
Bosačková, J. and Crane, R. K. (1965). *Biochim. biophys. Acta* **102**, 423.
Cinader, B. (ed.) (1967). *In* "Antibodies to Biologically Active Molecules", Proc. 2nd FEBS Meeting, Vienna, 1965, Vol. I, p. 85. Pergamon Press, Oxford.
Colombo, V., Balthazar, A.-K. and Semenza, G. (1969). Unpublished results.
Crane, R. K. (1962). *Fedn Proc. Fedn Am. Socs exp. Biol.* **21**, 891.
Csáky, T. Z. (1963). *Fedn Proc. Fedn Am. Socs exp. Biol.* **22**, 3.
Cummins, D. L., Gitzelmann, R., Lindenmann, J. and Semenza, G. (1968). *Biochim. biophys, Acta* **160**, 396.
Deutsch, H. F. (1967). *In* "Methods in Immunology and Immunochemistry" (C. A. Williams and M. W. Chase, eds.) Vol. 1, p. 315. Academic Press, London and New York.
Diedrich, D. F. (1966). *Archs Biochem. Biophys.* **117**, 248.
Diedrich, D. F. (1968). *Archs Biochem. Biophys.* **127**, 803.
Eichholz, A. (1969). *Fedn Proc. Fedn Am. Socs exp. Biol* **28**, 1, 30.
Fahey, J. L. (1967). *In* "Methods in Immunology and Immunochemistry" (C. A. Williams and M. W. Chase, eds.) Vol. 1, p. 321. Academic Press, London and New York.
Faust, R. G. (1964). *J. cell. comp. Physiol.* **63**, 119.
Faust, R. G., Leadbetter, M. G., Plenge, R. K. and McCaslin, A. J. (1968). *J. gen. Physiol.* **52**, 482.
Gitzelmann, R., Bachi, Th., Binz, H., Lindenmann, J. and Semenza, G. (1970). *Biochim. biophys. Acta.* **196**, 20.
Hamilton, J. D and McMichael, H. B. (1968). *Lancet* ii, 154.
Ito, S. (1969). *Fedn Proc. Fedn Am. Socs exp. Biol.* **28**, 1, 12.
Kolínská, J. and Semenza, G. (1967). *Biochim. biophys. Acta* **146**, 181.
Kotyk, A. (1970). This volume, p. 99.
Malathi, P. and Crane, R. K. (1969). *Biochim. biophys. Acta* **173**, 245.
Miller, D. and Crane, R. K. (1961). *Biochim. biophys. Acta* **52**, 281.
Newey, H., Sanford, P. A. and Smyth, D. H. (1965). *Nature, Lond.* **205**, 389.
Nishi, Y., Yoshida, T. O. and Takesue, Y. (1968). *J. molec. Biol.* **37**, 441.
Parsons, D. S. and Prichard, J. S. (1965). *Nature, Lond.* **208**, 1097.
Parsons, D. S. and Prichard, J. S. (1966). *Biochim. biophys. Acta* **126**, 471.
Prichard, J. S. (1969). *Nature, Lond.* **221**, 369.
Prichard, J. S. and Parsons, D. S. (1968). *J. Physiol., Lond.* **199**, 137.
Rutloff, H., Friese, R. and Täufel, K. (1965). *Hoppe-Seyler's Z. Physiol. Chem.* **341**, 134.
Semenza, G. (1967a). *Caries. Res.* **1**, 187.
Semenza, G. (1967b). *Protides biol. Fluids* **15**, 201.
Semenza, G. (1968). *In* "Handbook of Physiology", Section on Alimentary Canal, (C. F. Code, J. R. Brobeck, H. W. Davenport, M. I. Grossman, C. L. Prosser, T. H. Wilson and W. Heidel, eds.), Vol. V, p. 2543. Amer. Ass. Physiol., Washington.

Semenza, G. (1969a). *Biochim. biophys. Acta* **173**, 104.
Semenza, G. (1969b). *Eur. J. Biochem.* **8**, 518.
Semenza, G., Tosi, R. and Delachaux, M. C. (1963). *Helv. chim. Acta* **46**, 1765.
Semenza, G., Tosi, R., Vallotton-Delachaux, M. C. and Mülhaupt, E. (1964). *Biochim. biophys. Acta* **89**, 109.
Semenza, G., Auricchio, S., Rubino, A., Prader, A. and Welsh, J. D. (1965). *Biochim. biophys. Acta* **105**, 386.
Semenza, G., Curtius, C. H., Kolínská, J. and Müller, M. (1967). *Biochim. biophys. Acta.* **146**, 196.
Ugolev, A. M. (1968). *Die Nahrung* **10**, 484.

Effect of Phloretin and Phlorizin on Sugar and Amino Acid Transport Systems in Small Intestine*

F. ALVARADO

*Departments of Pharmacology and Physiology,
University of Puerto Rico School of Medicine,
San Juan, Puerto Rico*

The first step in transport across a biological membrane is generally believed to be a reversible combination between the substrate and a specific membrane receptor or carrier (Wilbrandt and Rosenberg, 1961). Specific receptors for the transport of sugars and amino acids are known to exist in many biological membranes, including the mucosal epithelial cell membrane (brush border) of the small intestine. However, in the hamster small intestine, recent evidence indicates that the receptors for sugars, neutral amino acids and basic amino acids, although different, are not entirely independent, but may be associated to form some kind of supramolecular structure having the kinetic properties of a "polyfunctional carrier" (Alvarado, 1970).

Other work also indicates that receptors for phenols may also occur in some membranes, particularly the red cell membrane (LeFevre, 1961). In the hamster small intestine one such phenol-binding site has been postulated to be involved in the allosteric, i.e. indirect, inhibition of sugar transport by the polyphenol, phloretin. Evidence has been presented that phloretin binds to a site different from, but very close to, the sugar-binding site; simultaneous binding of the phloretin-glucoside, phlorizin, to both the sugar- and the phloretin-binding site can account for the enormous potency of phlorizin as an inhibitor of sugar transport in hamster intestine (Alvarado, 1967).

Carrying the above reasoning a step further, the prediction may perhaps be made that, if both amino acids and phloretin act as allosteric inhibitors of sugar transport in hamster small intestine, it may be possible that phloretin also inhibits amino acid transport in the same manner. I am now exploring this possibility and will summarize here some of the results obtained.

The evidence indicating that sugars and amino acids act as allosteric inhibitors on each other's transport has been discussed recently (Alvarado, 1970), and will

---

* Contribution No. 10 of the Laboratory of Neurobiology. Supported in part by Research Career Development Award No. 1 KO4 AM 42383, and Research Grant No. 1 RO1 AM-13118 from the National Institute of Arthritis and Metabolic Diseases, U.S.P.H.S.

not be repeated here. Briefly, this evidence consists in the following: (a) sugar transport is inhibited by the amino acids, and vice versa; (b) this inhibition takes place from the external side of the membrane, and reflects a specific effect on carrier-mediated substrate entry; (c) this inhibition seems to follow typical kinetics of *pseudocompetitive inhibition* (also called *partially competitive inhibition*; see Alvarado, 1967), thus suggesting that the membrane binding sites involved in sugar and amino acid transport, respectively, albeit different, are closely associated and influence each other allosterically; and (d) sugars elicit countertransport of amino acids, and vice versa.

The question I posed then, at the beginning of the present studies was: does phloretin act on sugar and amino acid transport by a similar mechanism, i.e. through its specific binding to a phenol site allosterically associated with the sugar- and amino acid-binding sites mentioned above? The results thus far seem to indicate that, although phloretin may have some action of this nature, its overall effect on transport is much more complex.

As predicted, phloretin and other phenols were first found to inhibit both sugar and amino acid transport; similar observations have been made in other laboratories (Adamic and Bihler, 1967; Hand et al., 1966; Kotyk et al., 1965). These effects were further characterized by studying transport at constant substrate and increasing phloretin concentrations. Similar to the case of sugar transport, phloretin seemed unable to block amino acid transport completely; on the contrary, as the phenol concentration was increased, the inhibition seemed to approach a limiting value, as shown in Fig. 1 for the amino acid, cycloleucine (cycloleucine is used here as a model for the neutral amino acids). Also, the phloretin effect did not seem to occur indirectly through inhibition of energy-yielding reactions within the cell since, at the higher phloretin concentrations used, transport was still taking place against an apparent concentration gradient (Fig. 1). Furthermore, the phloretin effect seems to be specific. Although other phenols such as saligenin were also found to inhibit sugar and amino acid transport, it seems that they act by a different mechanism. For example, preliminary studies with saligenin indicate that this phenol is able to fully block amino acid and sugar transport, although only at high concentrations (above 70 mM). The nature of this inhibition is not clear, but it is possible that it may involve indirect effects on metabolism. What is more interesting at this moment is that saligenin and phloretin do not compete for the same binding site, as shown by the fact that their inhibitory effects are additive (Table 1).

If work had been stopped at this point, one might have been able to conclude that, indeed, phloretin is an allosteric inhibitor of both sugar and amino acid transport. More recent observations, however, reveal the complexity of action of this polyphenol. In the first place, and in contrast, for instance, with galactose (Alvarado, 1968), the inhibitory effect of phloretin on cycloleucine transport is time-dependent, i.e. it cannot be extrapolated to zero time (Fig. 2). This result,

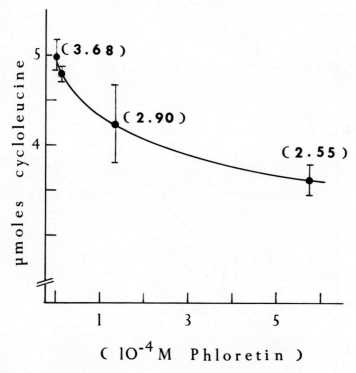

Figure 1. Effect of phloretin on cycloleucine transport in hamster small intestine. Transport was determined (tissue accumulation method) in Krebs-Henseleit phosphate buffer, pH 7.2, containing 1.5 mM [^{14}C]cycloleucine, and phloretin as indicated. Results are expressed as μmoles accumulated in 5 min per ml tissue water, corrected for the extracellular space. Figures in parentheses are the concentration ratios tissue/medium, where values greater than one indicate accumulation against the gradient. Each point is the mean of three to six determinations ± S.E.

Table 1. The additive effects of saligenin and phloretin as inhibitors of cycloleucine transport

Addition	Per cent inhibition	Tissue/medium concentration ratios
30 mM Mannitol	–	3.76
Mannitol + 0.5 mM phloretin	17.6	3.09
30 mM Saligenin	40.4	2.19
Saligenin + phloretin	62.2	1.39

Rings of intestine were incubated for 5 min in 5 ml phosphate buffer containing 1.5 mM [^{3}H]cycloleucine and the indicated effectors. To make conditions homogeneous, mannitol was used in the controls. Results are means of quadruplicate determinations, corrected for the extracellular space.

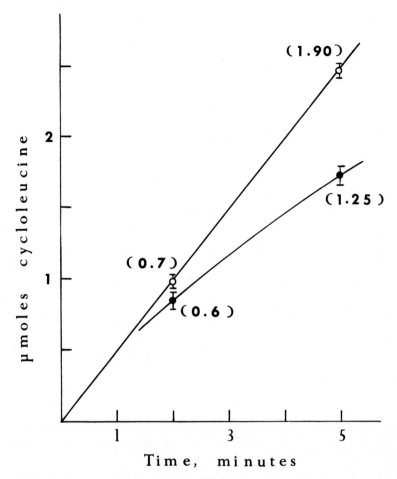

Figure 2. Time dependence of phloretin inhibition of cycloleucine transport. Same conditions as in Fig. 1. One series (○) was incubated in the absence and the other in the presence (●) of 0.6 mM phloretin. Points are means of eight separate determinations. Bars indicate ± S.E. Figures in parentheses are the concentration ratios tissue/medium.

however, is not necessarily contrary to the hypothesis of an allosteric effect of phloretin. It could indicate, for instance, that there are a large number of equivalent phloretin-binding sites in the membrane that are filled up progressively and at random; it is only when those sites associated with transport carriers are filled that amino acid (or sugar) transport is inhibited.

Such an interpretation seems somewhat speculative, but nevertheless is consistent with the experimental facts and is supported by other lines of evidence. In the first place, we have the work of LeFevre and Marshall (1959) on the effect of phloretin on sugar transport in red blood cells. They found that phloretin is tightly bound at the membrane surface, this binding being highly specific. More interestingly, as the phloretin concentration is increased (the limit to this is set by the low solubility of phloretin in water), there is "no evidence of the saturation behaviour anticipated by the presumption of a fixed number of sites of high affinity". From their data they concluded that phloretin binds to a large number of specific binding sites in the membrane (some intracellular binding may also occur, although the bulk of the attachment is at the cell surface), and "the sites of inhibition of the sugar transport must represent only a tiny fraction of the potential sites for phloretin attachment". It seems quite conceivable that an essentially identical situation occurs in the intestine, so that, as suggested above, phloretin would bind to a large number of membrane sites, some of which may be associated to sugar-binding sites, others to amino acid-binding sites, and perhaps others to both. The affinity of phloretin (and phlorizin) for the membrane may to a large extent be related to the known tendency of these compounds to attach themselves to proteins. For example, when treated with Somogyi's deproteinizing reagent (zinc sulfate plus barium hydroxide) or with trichloroacetic acid (another well-known deproteinizing reagent), phlorizin and phloretin remain in the supernatant. However, if protein is present during the treatment, both polyphenols will go with the protein to the precipitate. Most of the binding in this case may be non-specific; for instance, hydrophobic bonds may be largely involved.

These considerations lead one back to the hypothesis recently presented by myself to explain the mechanism of action of phlorizin (Alvarado, 1967). In this hypothesis, phlorizin is postulated to bind simultaneously to two different binding sites: one specific for its sugar moiety and one specific for the aglycone. What is the biochemical meaning of this latter site? Is it the same site at which phloretin must bind in order to inhibit sugar and amino acid transport? Not necessarily, although such identity is indicated by the lack of additivity of the effects of phlorizin and phloretin on sugar (Alvarado, 1967) and amino acid (Table 2) transport. The postulated phenol site is a kinetic site, i.e. it serves to explain why phlorizin exhibits an affinity for the sugar carrier three orders of magnitude higher than glucose and other free hexoses.

But, of course, both phlorizin and particularly its aglycone, phloretin, are highly lipophilic molecules that might be expected to bind to many points throughout the entire surface of the membrane. From this point of view, it cannot be said that there are any "specific" phenol-binding sites. At the same time, however, any membrane site at which phloretin will tend to bind, resulting at the same time in some effect on sugar and/or amino acid transport, may legitimately be considered a "specific" transport-modifying site. It is also conceivable that at least some of these sites may be coincident with substrate-binding sites, which may explain why, in some circumstances, inhibition by phloretin does not reach a plateau but may tend to fully block sugar and amino

Table 2. The effect of phlorizin and phloretin on the active transport of cycloleucine in small intestine

Condition	Transport rate	T/M
Control	4.49 ± 0.22	3.63
+ phlorizin	3.96 ± 0.14	3.00
+ phloretin	3.08 ± 0.15	2.28
+ phloretin + phlorizin	3.12 ± 0.18	2.33

Rings of hamster small intestine were incubated for 5 min in 5 ml phosphate buffer containing 1.5 mM [^{3}H]cycloleucine and the indicated effectors. Results are means of five separate experiments where the phlorizin concentration was 0.1 mM (0.2 mM in one experiment); the phloretin concentration ranged from 0.59 to 0.63 mM; and the pH also varied in the range 7.0 to 7.9 (conditions were kept invariant within a given experiment). Transport rates are indicated in μmoles of substrate transported in 5 min per ml tissue water, corrected for the extracellular space; each value given is the average of twenty determinations ± S.E. The T/M ratios represent the concentration ratios tissue/medium, where a value higher than 1.0 indicates active transport.

acid transport, as indicated by other observations (Alvarado, unpublished results) using long periods of preincubation with phloretin.

Before ending this brief review on phloretin and phlorizin action on intestinal transport, I would like to comment on some recent developments that have brought fresh interest to this already exciting problem. I refer to the recent demonstration of significant phlorizin hydrolase activity in hamster small intestine, accomplished independently and simultaneously by Diedrich (1968) in Lexington and by Malathi and Crane (1968) at Rutgers. As first pointed out by Diedrich, both isolated brush borders and pieces of intestinal tissue incubated with phlorizin in conditions similar to those used for studying intestinal transport, rapidly bind the glucoside, although some of this is hydrolysed and appears both in the medium and in bound form as the aglycone, phloretin. Phloretin formation and binding to isolated brush borders occurs quite rapidly, so that, within the first minute of incubation, the concentration ratio phloretin:phlorizin is about 1 and increases to 2.75 and 4.0 at one and two hours, respectively (calculated from the data in Fig. 6, Diedrich, 1968).

It is essentially on the basis of this evidence that Diedrich has suggested the possibility that "it is phloretin and not the glucoside which actually inhibits intestinal sugar transport". As an extension of this suggestion, Diedrich also questions the validity of available quantitative data on phlorizin inhibition "since it is essentially all phloretin which is bound to the intestinal epithelia, and not the parent glucoside". These are challenging suggestions. They could perhaps mean that a thorough review of our views on the mechanism of action of phlorizin is urgently needed. At the same time, there are data in the literature, including Diedrich's own data, that would seem to argue against what might be called "the phloretin or indirect hypothesis" of phlorizin inhibition. One well established fact about phlorizin is that it acts instantly, i.e. a low but appropriate concentration of this compound (e.g. 10^{-4}M) fully blocks sugar transport, and this inhibition may be extrapolated to zero time. In agreement with this, it seems apparent that the amount of phlorizin bound to isolated brush borders (see Fig. 6, Diedrich, 1968) is constant for at least 120 min, maximal phlorizin binding being attained immediately after incubation is begun (the phlorizin binding curve is a horizontal line). On the contrary (see the same figure), the amount of phloretin bound to brush borders increases with time, and it seems that the phloretin binding curve can indeed be extrapolated to zero time. The conclusion that may be drawn from these considerations is obvious. Regardless of the amounts of phloretin that may be bound to brush borders, a constant proportion of bound phlorizin is always present, and this amount of phlorizin can fully account for the inhibition of sugar transport.

The same reasoning may be applied to the case of the effective concentration of phlorizin in the incubation medium. Although some phlorizin is undoubtedly hydrolysed, this hydrolysis is by no means instantaneous or complete. In Diedrich's own experiments, after incubations as long as 120 min, nearly 60% of the total phenol in the medium is still phlorizin. This would indicate that the error due to phlorizin hydrolysis in the calculation of kinetic parameters of phlorizin inhibition may be as little as a factor of two or less (probably much less since most kinetic measurements are done at considerably shorter time intervals). It is also known that presence of phloretin, even in a 70-fold excess over phlorizin, does not influence the effect of the glucoside on sugar transport (Alvarado, 1967).

Another argument one can adduce against the hypothesis of the indirect mechanism of action of phlorizin is the difference between phlorizin and phloretin with regard to reversibility. Ponz and Lluch (1955) were the first to clearly show that the inhibition caused by phlorizin is freely reversible *in vivo*, a finding that has been amply confirmed, both *in vivo* and *in vitro*, using intestine from several animal species. The free reversibility of the effect of phlorizin, together with the instantaneous nature of the inhibition, constitutes a strong indication that phlorizin is indeed a fully competitive inhibitor of sugar transport, as shown by more recent kinetic experiments (Alvarado, 1967).

The situation with phloretin is entirely different. In contrast with phlorizin, the onset of phloretin inhibition requires time, as shown before (Fig. 2), and the inhibited curve cannot be extrapolated to zero time, i.e. phloretin inhibition is not instantaneous, as is the case with phlorizin, a true fully-competitive inhibitor. Furthermore, the inhibition by phloretin is not reversible, as is illustrated in Table 3. If phloretin inhibition is not freely reversible, this argues strongly against the view that phlorizin, a readily reversible inhibitor, acts through local formation of phloretin.

Table 3. Lack of reversibility of the effect of phloretin on sugar transport

	1	2	3	4
Phloretin in first incubation	+	+	0	0
Phloretin in second incubation	+	0	+	0
Glucoside transported (μmoles per ml tissue water)	1.00	1.65	1.63	3.44
T/M ratios	0.67	1.15	1.12	2.58

Rings of hamster intestine were first incubated for 15 min in 5 ml phosphate buffer containing, as indicated, 0.64 mM phloretin. After this preincubation, transport of methyl-α-glucoside was measured by transferring to and reincubating the tissues for 5 min in 5 ml buffer containing 1.15 mM substrate and, as indicated, 0.615 mM phloretin. Results are expressed as μmoles of glucoside accumulated per ml tissue water, corrected for the extracellular space; and as the concentration ratios tissue/medium (T/M). Each point is the mean of a quadruplicate determination.

In conclusion, in the brush borders there are specific phlorizin-binding sites composed of two separate parts: one glucose-binding site and one phloretin-binding site. The glucose-binding site is the same involved in sugar active transport through this membrane. Binding of the glucose moiety of phlorizin to this glucose-binding site explains why phlorizin behaves as an authentic fully-competitive inhibitor for sugar transport. Simultaneous binding of the phloretin moiety of phlorizin to the phloretin-binding site explains why the overall affinity of phlorizin for the sugar carrier is three orders of magnitude higher than that exhibited by the free sugars.

Phlorizin bound to its specific (double) binding site remains intact. With time, some phlorizin may be hydrolyzed at other points on the membrane: part of the resulting phloretin may go back to the medium and part may remain bound to phloretin-binding sites throughout the surface of the membrane. These sites may be similar to some of those involved in phlorizin binding. The difference is that the phlorizin-selective sites are associated with sugar-binding sites: this spacial association with sugar-binding sites is precisely what makes this particular family of phloretin-binding sites unique.

REFERENCES

Adamic, S. and Bihler, I. (1967). *Molec. Pharmacol.* **3**, 188.
Alvarado, F. (1967). *Biochim. biophys. Acta* **135**, 483.
Alvarado, F. (1968). *Nature, Lond.,* **219**, 276.
Alvarado, F. (1970). In "Intestinal Transport of Electrolytes, Amino Acids, and Sugars" (W. McD. Armstrong and A. S. Nunn, Jr., eds.). Charles C. Thomas, Springfield, Illinois, in press.
Diedrich, D. F. (1968). *Archs. Biochem Biophys.* **127**, 803.

Hand, D. W., Sanford, P. A. and Smyth, D. H. (1966). *Nature, Lond.,* **209**, 618.
Kotyk, A., Kolinska, J., Veres, K. and Szammer, J. (1965). *Biochem. Z.* **342**, 129.
LeFevre, P. G. (1961). *Pharmac. Rev.* **13**, 39.
LeFevre, P. G. and Marshall, J. K. (1959). *J. biol. Chem.* **234**, 3022.
Malathi, P. and Crane, R. K. (1968). *Fedn Proc. Fedn Am. Socs exp. Biol.* **27**, 385.
Ponz, F. and Lluch, M. (1955). *Rev. esp. Fisiol.* **11**, 267.
Wilbrandt, W. and Rosenberg, T. (1961). *Pharmac. Rev.* **13**, 109.

Author Index

Numbers followed by an asterisk refer to the page on which the reference is listed.

A

Adamic, S., 132, 138*
Algaranti, I. D., 48, 48*
Alvarado, F., 123, 125, 129*, 131, 132, 135, 137, 138*
Anderson, B. E., 87, 88, 90, 95, 96*, 97*, 103, 105, 106*, 107*
Anderson, C. M., 118, 129*
Anderson-Cedegren, E., 17, 21, 32*
Anderson, R. L. 87, 96*
Anraku, Y., 101, 103, 106*
Asai, H., 11, 16*
Auricchio, S., 118, 129*, 130*
Azam, F., 101, 103, 107*

B

Baarda, J. R., 73, 76, 79*
Balthazar, A-K., 118, 125, 126, 129*
Behrens, N. H., 48, 48*
Benedetti, E. L., 11, 14, 15, 16*
Benson, A. A., 36, 37*
Bernheim, F., 28, 32*
Bihler, I., 132, 138*
Biryuzova, V. I., 57, 58*
Blair, P. V., 21, 32*
Bobinski, H., 101, 102, 106*
Bonsall, R. B., 101, 103, 106*
Bosacková, J., 118, 129*
Boulton, A. A., 39, 42, 44, 45, 48*
Branton, D., 4, 7*
Braun, P. E., 48, 48*
Briscas, E., 61, 63, 65, 70*
Buttin, G., 87, 97*

C

Cabib, E., 48, 48*
Caminatti, H., 48, 48*
Campbell, J. N., 62, 69*
Carlson, H. E., 28, 32*
Carter, J. R., 101, 103, 106*
Cash, W. D., 28, 32*
Caspary, W. F., 111, 115, 115*
Chance, B., 17, 22, 32*
Chapman, D., 5, 7*
Chappell, J. B., 73, 78*, 79*
Christensen, H. N., 83, 84, 84*, 85*
Christophersen, B. O., 28, 32*
Cinader, B., 125, 129*
Cirillo, V. P., 105, 106*
Cohen, G. N., 87, 93, 96*, 97*
Coleman, R., 5, 7*, 11, 16*
Colombo, V., 118, 125, 126, 129*
Crane, R. K., 110, 111, 112, 114, 115, 115*, 116*, 117, 118, 119, 120, 125, 129*, 136, 139*
Crofts, A. R., 73, 78*
Csáky, T. Z., 128, 129*
Cullen, A. M., 84, 84*
Cummins, D. L., 125, 129*
Curtius, C. H., 118, 130*

D

Danielli, J. F., 3, 7*
Davidson, M., 118, 129*
Davson, H., 3, 7*
Dedonder, R., 95, 97*
Deenen, L. L. M., van, 12, 16*
Delachaux, M. C., 120, 130*
Deutsch, H. F., 125, 129*
Dezelee, Ph., 61, 63, 65, 70*
Diedrich, D. F., 125, 129*, 136, 137, 138*
Dourmashkin, R. R., 14, 16*
Dreyfus, J., 101, 107*
Dreyer, W. J., 102, 103, 106*

E

Eagan, J. B., 87, 95, 96*, 105, 106*

Eddy, A. A., 47, 48*
Eichholz, A. 128, 129*
Ekong, E. A., 28, 32*
Elbaz, L., 61, 63, 65, 70*
Ellar, D. J., 52, 53, 54, 55, 57, 58*
Emmelot, P., 11, 14, 15, 16*
Englesberg, E., 87, 93, 95, 96*

F

Fahey, J. L., 125, 129*
Fairclough, G. F., Jr., 102, 106*
Falcoz-Kelly, F., 95, 96*
Faust, R. G., 119, 128, 129*
Fernandez-Morán, H., 18, 21, 32*
Finean, J. B., 5, 7*, 9, 11, 12, 16*
Fink, J., 21, 28, 32*
Fleischer, B., 4, 7*
Fleischer, S., 4, 7*
Ford, L., 28, 32*
Fortney, S. R., 28, 32*
Fox, C. F., 101, 103, 106*
Fraenkel, D. G., 95, 96*, 97*
Freeman, M., 118, 129*
Freer, J. H., 52, 53, 54, 57, 58*
Frezal, J., 118, 129*
Friese, R., 117, 129*
Fruton, J. S., 102, 106*

G

Gachelin, G., 89, 90, 92, 96*
Garcia Lopez, M. D., 47, 48*
Garcia Mendoza, C. G., 39, 42, 44, 46, 47, 48*
Gardy, M., 28, 32*
Gascón, S., 45, 46, 48*
Gebicki, J. M., 28, 32*
Gelman, N. S., 57, 58*
Ghosh, S., 87, 97*, 105, 107*
Ghuysen, J. M., 61, 62, 63, 65, 69*
Gitzelmann, R., 125, 128, 129*
Gordon, A. S., 12, 16*
Gorter, E., 3, 7*
Green, D. E., 21, 32*
Green, W. A., 5, 7*
Grendel, F., 3, 7*
Guerra, F., 18, 21, 28, 32*
Guinand, M., 63, 69*

H

Haarhoff, K. N., 73, 79*
Hamilton, J. D., 119, 129*
Hamilton, W. A., 71, 77, 79*
Hand, D. W., 132, 139*
Handlogten, M. E., 83, 84, 84*, 85*
Hanson, T. E., 87, 96*
Harold, F. M., 73, 76, 79*
Harris, E. J., 115, 116*
Hartman, P. E., 87, 88, 95, 97*, 103, 105, 107*
Haskovec, C., 100, 103, 106*
Heijenoort, J., van, 61, 63, 65, 70*
Hengstenberg, W., 87, 96*
Heppel, L. A., 90, 97*, 101, 106*
Herz, F., 11, 16*
Höfer, M., 71, 79*
Hoffee, P. A., 87, 93, 95, 96*
Hoffsten, P. E., 28, 32*
Hopfer, V., 76, 79*
Horecker, B. L., 95, 96*
Hotchkiss, R. D., 71, 79*
Huet, C., 30, 32*
Huff, J. W., 28, 32*
Hummel, J. P., 102, 103, 106*
Hunt, S., 101, 103, 106*
Hunter, F. E., Jr., 21, 28, 32*
Hurwitz, A., 21, 32*

I

Inesi, G., 11, 16*
Ito, S., 119, 129*
Izaki, K., 65, 68, 69*

J

Jagger, W. S., 115, 116*
Johnson, J. H., 115, 116*
Jones, T. H. D., 101, 106*
Jos, J., 118, 129*

K

Kaback, H. R., 78, 79*, 90, 96*
Kandler, O., 63, 69*
Kaplan, E., 11, 16*
Karlsson, U., 17, 21, 32*
Kates, M., 51, 58*
Kennedy, E. P., 101, 103, 106*

Kepes, A., 87, 88, 93, 96*, 97*, 103, 105, 106*
Kessler, R., 87, 93, 97*
Knights, B. A., 39, 42, 44, 48*
Kolber, A. R., 99, 100, 103, 106*
Kolinská, J., 118, 120, 121, 129*, 130*, 132, 139*
Korn, E. D., 3, 4, 5, 7*
Kotyk, A., 100, 101, 103, 105, 106*, 107*, 117, 122, 125, 129*, 132, 139*
Kundig, F. D., 87, 88, 90, 95, 96, 97*, 103, 105, 107*
Kundig, W., 87, 88, 90, 95, 96, 96*, 97*, 103, 105, 106*, 107*

L

Lam, I., 84, 85*
Lampen, O. J., 44, 45, 46, 47, 48*
Lamy, F., 87, 96*
Leadbetter, M. G., 128, 129*
Le Fevre, P. G., 131, 134, 139*
Lehninger, A. L., 76, 79*
Lekim, D., 48, 49*
Lenard, J., 12, 16*
Lepesant, J. A., 94, 97*
Levine, M., 105, 107*
Levinthal, M., 87, 88, 95, 97*, 103, 105, 107*
Levy, J. F., 21, 32*
Levy, M., 30, 32*
Leyh-Bouille, M., 62, 69*
Limbrick, A. R., 5, 7*
Lin, E. C. C., 88, 95, 97*
Lindenmann, J., 125, 129*
Lluch, M., 137, 139*
Longley, R. P., 39, 42, 44, 48*
Longton, J. J., 47, 48*
Lukoyanova, M. A., 57, 58*
Lynn, W. S., Jr., 28, 32*
Lyon, I., 112, 116*

M

Macarulla, J. M., 18, 21, 32*
MacBrinn, M. C., 37, 37*
McCaslin, A. J., 128, 129*
McMichael, H. B., 119, 129*
Malathi, P., 125, 129*, 136, 139*
Marchesi, V. T., 10, 16*

Marshall, J. K., 134, 139*
Martonosi, A., 11, 12, 14, 15, 16*
Matile, Ph., 39, 40, 42, 44, 45, 46, 48*
Matsuhashi, M., 65, 70*
Messer, M., 118, 129*
Miller, D., 117, 118, 119, 129*
Mitchell, P., 76, 78, 79*
Mitchell, R. F., 17, 21, 32*
Monod, J., 87, 93, 96*, 97*
Montenecourt, B. S., 45, 48*
Moor, H., 4, 7*, 39, 40, 45, 46, 48*
Moore, C., 73, 79*
Morgan, H. E., 105, 107*
Morse, M. L., 87, 95, 96*, 105, 106*
Moyle, J., 76, 79*
Muhlethaler, K., 4, 7*, 39, 40, 45, 46, 48*
Mülhaupt, E., 120, 130*
Müller, M., 118, 130*
Munoz, E., 52, 53, 54, 55, 56, 57, 58*

N

Nachbar, M. S., 52, 53, 54, 58*
Neu, M. C., 90, 97*
Neumann, N. P., 45, 48*
Newey, H., 123, 129*
Ng, M. H., 52, 53, 54, 56, 58*
Nichoalds, G. E., 101, 107*
Nishi, Y., 128, 129*
Nurminen, T., 39, 42, 49*

O

O'Brien, J. S., 33, 36, 37, 37*, 38*
Oda, T., 21, 32*
Ottolenghi, A., 28, 32*
Ottolenghi, P., 45, 48*
Oura, E., 39, 42, 49*
Oxender, D. L., 101, 105, 107*

P

Palade, G. E., 10, 16*
Pardee, A. B., 101, 104, 105, 107*
Parsons, D. F., 17, 18, 22, 32*
Parsons, D. S., 117, 118, 122, 129*
Pascaud, M., 30, 32*
Penrose, W. R., 101, 107*
Perkins, H. R., 62, 69, 69*
Petit, J. F., 61, 63, 65, 70*

Piperno, J. R., 101, 107*
Plenge, R. K., 128, 129*
Pogell, B. M., 96, 97*
Ponz, F., 137, 139*
Prader, A., 118, 129*, 130*
Pressman, B. C., 71, 73, 79*, 115, 116*
Prestidge, L. S., 101, 107*
Prichard, J. S., 117, 118, 119, 122, 129*

R

Radojkovic, J., 103, 107*
Regen, D. M., 105, 107*
Rey, J., 118, 129*
Rickenberg, H. W., 87, 97*
Robertson, J. D., 3, 7*
Rose, A. H., 39, 42, 44, 48*
Roseman, S., 87, 88, 90, 95, 96, 96*, 97*, 103, 105, 106*, 107*
Rosenberg, Th., 103, 105, 107*, 131, 139*
Rosse, W. F., 14, 16*
Rotman, B., 103, 107*
Rubino, A., 118, 129*, 130*
Rutloff, H., 117, 129*

S

Salton, M. R. J., 6, 51, 52, 53, 54, 55, 56, 57, 58*, 71, 79*
Sampson, E. L., 33, 37, 38*
Sanford, P. A., 123, 129*, 132, 139*
Santiago, E., 18, 21, 32*
Schatzman, H. J., 15, 16*
Schauenstein, E., 30, 32*
Schleifer, K. H., 63, 69*
Schmidt, 6
Schneider, A., 28, 32*
Schor, M. T., 52, 53, 54, 56, 58*
Schrier, S. L., 16*
Schutz, B., 28, 32*
Scott, A., 28, 32*
Semenza, G., 118, 119, 120, 121, 125, 126, 128, 129*, 130*
Shapiro, O. W., 28, 32*
Simakova, I. M., 57, 58*
Simoni, R. D., 87, 88, 95, 96*, 97*, 103, 105, 106*, 107*
Singer, S. J., 12, 16*

Sjostrand, F. S., 3, 7*, 17, 21, 32*
Skrede, S., 28, 32*
Smith, E., 28, 32*
Smith, M. F., 88, 97*
Smyth, D. H., 123, 129*, 132, 139*
Snell, E. E., 48, 48*
Stadtman, E. R., 78, 79*
Stahl, E., 44, 49*
Stein, W. D., 99, 100, 101, 102, 103, 105, 106*, 107*, 114, 116*
Steveninck, J., van, 105, 107*
Stevenson, J. H., 11, 16*
Stevenson, N. R., 115, 115*
Sticht, G., 48, 49*
Stoeckenius, W., 4, 7*, 21, 32*
Stoffel, W., 48, 49*
Straus, J. H., 12, 16*
Strominger, J. L., 65, 68, 69, 69*
Suomalainen, H., 39, 42, 49*
Szammer, J., 132, 139*

T

Tager, H. S., 84, 85*
Takesue, Y., 128, 129*
Tanaka, S., 88, 95, 97*
Täufel, K., 117, 129*
Thiele, E. H., 28, 32*
Thomas, E. L., 83, 85*
Thompson, J. E., 11, 16*
Thompson, T. E., 76, 79*
Tipper, D. J., 69, 69*
Tosi, R., 120, 130*
Townley, R. R. W., 118, 129*

U

Ugolev, A. M., 117, 130*
Uruburu, F., 47, 48*

V

Vallotton-Delachaux, M. C., 120, 130*
Vázquez, J., 18, 32*
Veres, K., 132, 139*
Villanueva, J. R., 39, 42, 44, 46, 47, 48*

W

Wagner, H., 44, 49*
Wallach, D. F. H., 12, 16*

Watanabe, K., 104, 107*
Watson, J. A., 93, 95, 96*
Weinstein, J., 28, 32*
Weisman, R. A., 4, 7*
Welsh, J. D., 118, 130*
Whippele, M. D., 101, 107*
Wilbrandt, W., 103, 105, 107*, 131, 139*
Wilbur, K. M., 28, 32*
Williams, G. R., 17, 22, 32*

Wills, E. D., 28, 32*
Wilson, H., 103, 107*
Winkler, H. H., 95, 96, 97*, 103, 107*

Y

Yoshida, T. O., 128, 129*

Z

Zand, R., 84, 85*
Zofcsik, W., 44, 49*

Subject Index

A

N-Acetyl glucosamine,
 peptidoglycans, in, 59, 61, 62
 transport of, 95
N-Acetyl mannosamine, 95
N-Acetyl muramic acid, in peptidoglycan, 59, 61, 62
Active transport, see also Transport, 81–139
Adenosine triphosphatase, (ATPase), of membranes 5, 11, 15, 45, 51–54
Aerobacter aerogenes, sugar transport in, 95, 104
D-Alanine carboxypeptidase, 65–69
Alveolar membrane, 1
Amino acids, see also under specific names, transport of,
 cationic, 82–83
 galactose and, 123
 phlorizin and, 123, 131–139
 sugar inhibition of, 123, 132
2-Aminobicydo(2,2,1)heptane-2-carboxylic acid, 84
Amphotericin, 44
Anaesthetics, 1
1,5-Anhydro-D-glucitol transport, 111, 112, 113
Antibacterials, see also under specific names,
 membrane active, 71–79
Antibiotics, see also under specific names,
 inactivation of, 59
 polypeptide, 71
Antibodies, sugar transport and, 125, 127
L-Arabinose transport protein, 102
Arbutin, 123, 124
Ascorbate, and mitochondrial membranes, 21–26, 28–30, 31–32
ATPase, see Adenosine triphosphatase

B

Bacillaceae, peptidoglycan of, 63
Bacillus megaterium,
 cetyl trimethylammonium bromide effect on, 77
 membrane-active antibacterials, and, 72–76
 peptidoglycan of, 63
B. subtilis,
 sugar transport in, 94
 wall peptide of, 66
Bacterial membranes,
 antibacterials, and, 71–79
 components of, 5, 6, 51–58
 enzymes in, 63–64
 transport systems of, 87–97, 99–107
 wall peptidoglycan, and, 59–69
Bacterial wall, membrane and, 59–69, 77
Bakers yeast, see Saccharomyces cerevisiae
Bimolecular leaflet model, of membranes, 3, 5, 33
Biogenesis, of membranes, 6–7
Biomembranes, see under specific names
Blood group substances, 4–5
Brush border membranes, sugar transport and, 109–116, 117–130
Butyribacterium rettgeri, peptidoglycan of, 63, 67

C

Candida utilis, plasma membrane of, 42, 44
Carbohydrates, see under specific names
Carrier systems, see also under Transport,
 for sugar transport, 114–115

Carrier theory, 110, 111
Cell membranes, see also under Membranes,
  bacterial wall in relation to, 59–70
  biochemical and structural studies of, 9–16
  chemical composition of, 5–6
  electron microscopy of, 4, 9, 10–14
  enzyme modification of, 10–16
  Micrococcus lysodeikticus of, 51–58
  molecular anatomy of, 33–38
  phospholipase C modification of, 12–16
  proteins of, 5, 6, 10
  structure of, 1–79
  transport mechanisms in, 81–139
  trypsin modification of, 10–12
Cell wall,
  degradation for membrane isolation, 51
  membrane support functions of, 59–69, 77
  yeast, plasmalemma and, 39, 40, 47, 48
Cephalothin, 69
Cetyl trimethylammonium bromide, (CTAB),
  bacterial membranes, and, 71, 72, 73
  bacteriostatic action of, 72, 74, 76
  inhibitory concentrations of, 72
  leakage from bacteria, 76, 77
  protoplast lysis by, 73, 76
  uptake of, 72, 77
Chemiosmotic hypothesis, 78
Chloropicrin, 1
Cholesterol, 5, 12, 15, 34, 36
Cholinesterase, 11
Circular dichroism, and membrane structure, 5, 9, 12
Conformational rearrangements, in transport, 110, 114–115
Corynebacterium, peptidoglycan, 63
C. poinsettiae, wall peptide, 60, 66, 67
Cristae mitochondriales, 18, 30, 31
CTAB, see Cetyl trimethylammonium bromide
Cycloleucine transport, phloretin and, 132, 133, 134
Cysteine, effect on mitochondrial membranes, 26–28, 28–32
Cytochromes, of bacterial membranes, 52–57
Cytoskeleton, 2

D

Danielli–Davson model, 2, 4
Demyelinating diseases, 37
Deoxycholate, (DOC), and membranes, 46, 47, 53
6-Deoxy-L-galactose, (6-DLGal), 110–114, 124
6-Deoxy-D-galactose, 124
1-DG, see 1-Deoxyglucose
1-Deoxyglucose, (1-DG), see also 1,5-Anhydro-D-glucitol
2-Deoxy-D-glucose, and glucose-fixation, 111, 112, 113, 123, 124
"2-Deoxy-D-glucose space", 120, 121
3-Deoxy-D-glucose, and glucose-fixation, 123, 124
6-Deoxy-D-glucose,
  glucose-fixation, and, 123, 124
  phlorizin and uptake of, 125 126
  uptake, antibodies, and, 127
Dinitrofluorobenzene, and glucose transport, 100–101
2,4,-Dinitrophenol, 93
Disassembly of mitochondrial membrane, 18–32
6-DLGal, see 6-Deoxy-L-galactose
DOC, see Deoxycholate
Double labelling technique, for membrane carriers, 99–100

E

EDTA, see Ethylene-diamine-tetra-acetic acid
Electron microscopy of membranes, 4, 9
  enzyme pretreatment, and, 10–11, 12–14
  mitochondria from, 19–20, 22, 25
Electron transport components, 51
Electrophoresis, of membrane proteins, 6

SUBJECT INDEX

Enzymes, *see also under specific names*,
  membrane modification by, 10–16
  Micrococcus membranes, of, 51–56
  yeast plasmalemma, of, 44–46
Equilibrium dialysis, 102
Ergosterol, 40, 42, 44
Erythrocyte ghosts,
  electron microscopy of, 12, 13, 14
  phospholipase C modification of, 12–16
  trypsin modification of, 10–12
  X-ray diffraction studies of, 5, 14–15
Erythrocyte membranes, 4–5, 6, 12–14, 15
Erythrocytes, glucose transport protein of, 100, 103
Escherichia coli,
  D-alanine carboxypeptidase of, 68
  β-galactoside permease of, 88, 95–96, 99–100
  membrane-active bacterials, and, 72–78
  peptidoglycan of, 61, 63, 64, 65
  sugar transport in, 87–97, 105
  transport proteins from, 99, 100, 101, 102, 103
N-Ethyl maleimide, (NEM), 90, 91, 101
Ethylene-diamine-tetra-acetic acid, (EDTA), 30

F

Ferritin-labelled antibody, 4
Freeze-etching technique, 4, 9, 40
Fructose transport, 95
D-Fructose, and glucose-fixation, 123, 124
D-Fucose, glucose-fixation and, 123
L-Fucose, 110

G

Gaffkya, peptidoglycan of, 61
Galactose,
  glucose-binding, and, 123, 124
  transport protein for, 100, 101, 103
β-Galactosidase, 99

β-Galactoside, transport of, 99–100, 103
β-Galactoside permease, 88
  galactoside transport, and, 95–96
  isolation of, 99–100
Glucose,
  transmural potential, and, 112, 113
  transport, 99, 100, 103, 104, 111
  dinitrofluorobenzene effect on, 100
  protein for, 103, 104
Glucose-binding protein, 101
Glucose-fixation, by enzyme complexes, 120–128
L-Glucose, 111, 112, 113
Glucose-6-phosphate, transport protein, 102
Glycerol transport, 95
Glycolipids, 47
GRAM, *see* Gramicidin
Gram negative bacteria, *see also under specific names*,
  membrane-active antibacterials, and, 76
  peptidoglycan structure of, 63
Gram positive bacteria, *see also under specific names*,
  compounds bacteriostatic to, 71–76
  peptidoglycan structure of, 63
Gramicidin, (GRAM), 71–76

H

Helical configuration of membrane proteins, 5, 9
Hexose phosphohydrolase, 89
Hpr (heat-stable protein) system, in sugar transport, 87–88, 90–91, 92, 96, 105
Hydrophobic bonding in membrane structure, 5

I

Infrared spectroscopy, membrane proteins and, 5
m-Inositol, and glucose fixation, 124
Insulin, 81
Intestinal sugar transport, *see under* Sugar transport

Invertase, yeast plasmalemma, 45, 46
Ion fluxes, 78
Ion permeability, antibacterials and, 73–76

L

Lactobacillus, peptidoglycan, 61, 68
L. acidophilus, 66, 68
Lamellae of mitochondria, 18–32
L-Leucine, transport protein for, 101, 103, 104
Leuconostoc, peptidoglycan, 61
Lipid cycle, in peptidoglycan biosynthesis, 63, 64, 65
Lipid, of membranes, 4, 5, 6, 9, 10, 11, 12–16, 23, 42, 44
Lipid peroxides, 28, 30
Lipoprotein of membranes, 6, 9, 11, 36, 43

M

M-protein, 101
Maltose transport, 95
Mannan, 45, 46, 47
Mannanprotein, 40, 45, 47–48
Mannitol transport, 95, 102
D-Mannitol, transmural potential with, 112, 113
Mannose transport, 95
D-Mannose, and glucose-fixation, 123, 124
Melibiose transport, 95
Membrane(s), *see also under* Cell membranes *and more specific names*,
  antibacterials, and, 71–79
  bacterial, *see also* Bacterial membranes
  components of, 51–58
  enzymes of, 63–64, 65
  transport across, 87–97, 99–107
  wall structures, and, 59–69
  biochemical and structural studies of, 9–16
  carriers, *see also* Transport proteins, 99–107
  diseases, and, 37
  enzymes of, 63–64, 65
  enzymic modification of, 10–16
  mitochondrial, 17–32

Membrane(s)–*continued*
  molecular anatomy of, 34–38
  negative staining of, 11, 12, 18–22, 31
  proteins of, 5, 6
  structural and biochemical studies of, 9–16
  structure of, 1–79
  transport mechanisms, and, 81–139
  transport proteins, *see also under* Transport, 99–107
  trilamellar structure, 5, 10, 11, 12
  yeast cytoplasmic, 39–49
Membrane-active antibacterials, 71–79
Membrane-specific enzymes, 4
Membranopathies, 37
Metachromatic leucodystrophy, 37
Methyl-D-glucopyranose, 123
Methyl-D-glucopyranoside, 124
3-O-Methyl-D-glucose, (3-MG),
  glucose-fixation, and, 123, 124
  intestinal transport of, 111, 112, 114
  uptake of sucrase, and, 120, 121
Methylglucoside transport, by *E.coli*, 81, 87–97
Methylglucoside-6-phosphate, 91, 92, 94
α-MG, *see* α-Methylglucoside
3-MG, *see* 3-O-Methyl-D-glucose
MIC, *see* Minimum inhibitory concentration
Micrococcus sp., peptidoglycan, 61, 62
M. lysodeikticus,
  cell membranes of, 6, 51–58
  membrane-active antibacterials, 72, 76, 77
  peptidoglycan of, 62
M. varians peptidoglycan, 63
Microsomes, 10, 11, 12, 14, 16
Minimum inhibitory concentration, of antibacterials, 72
Mitochondria,
  inner membrane of, 18–32
  ascorbate effect on, 21–26, 28–32
  cysteine effect on, 26–28, 28–32
  disassembly of, 18–32
  electron micrographs of, 19, 20, 22, 25
  thyroxine effect on, 28

Mitochondria—*continued*
 ultrastructure of, 17–21
 outer membrane of, 4
 structure of, diagram of, 31
Mobile carrier theory, 110
Muscle microsomes, 10, 11, 12, 14, 16
Mycoplasma group of membranes, 5
Myelin, 5, 6, 33, 34, 36, 37

N

Na^+, *see also* Sodium
Na^+, pump, 81
Na^+-dependent glucose-binding, 122, 123, 128
Na^+-dependent sugar transport, 110–115, 120
$NADH_2$-dehydrogenase, of membranes, 52, 53
 solubilization of, 54–55
Negative staining, of membranes, 11, 12, 18–22, 31
NEM, *see* N-Ethyl maleimide
NMR, *see* Nuclear Magnetic Resonance
Nuclear magnetic resonance, membrane structure and, 5, 9
Nucleic acids, and yeast plasmalemma, 43, 44
Nystatin, 44

O

Optical rotatory dispersion, membrane structure and, 5, 9, 12
ORD, *see* Optical rotatory dispersion
Osmium tetroxide, 3, 4, 21
Overton-Meyer theory, 1, 2
Oxacillin, 69
Oxidative phosphorylation, 71, 78

P

Papain, effect on sugar uptake, 82, 120, 121
Penicillin G, 69
PEP, *see* Phosphoenolpyruvate
Peptidoglycan, of bacterial wall, 59–69
 biosynthesis of, 59, 63–69
 membrane, and, 59–69
 structure of, 59–63
 types I, II, III and IV, 60–62

Permeases, *see also under name of transported substrate*,
 β-galactoside, for, 88, 95–96, 99–100
 α-methylglucoside, for, 88
 transport proteins, and, 87
Peroxidase-antibody method, for membrane proteins, 4
Phloretin, and transport, 131–139
Phlorizin,
 amino–transport, and, 131–139
 sugar transport, and, 82, 111, 113, 123, 124, 125, 126, 131–139
Phlorizin hydrolase, 125, 126, 136
Phosphatidyl choline, in mitochondrial membranes,
 ascorbate effect on, 23, 24, 31
 cysteine effect on, 27, 31
 thyroxine effect on, 28, 30, 31
Phosphatidyl cholinephosphate hydrolase, *see* Phospholipase C
Phosphatidyl ethanolamine, in mitochondrial membranes, 23, 24, 27, 28, 31
Phosphoenolpyruvate, and methylglucoside uptake, 87–97
Phosphoenolpyruvate-hexose phosphotransferase, 87–88, 89, 90, 91
Phospholipase C, 10, 12–16
Phospholipids, *see also under specific names*,
 mitochondrial, 21–32
 phospholipase C, and, 12–16
 yeast plasma membranes of, 42
Phosphorylative sugar transport, 87–97, 105
Phosphotransferases, 87–88, 105
Plasma membranes, *see also* Membranes *and under more specific names*,
 ATPase of, 15
 enzymic modification of, 11, 12–16
 yeast, of, 39–49
Plasmalemma, *see under* Yeast
Polypeptide antibiotics, *see also under specific names*, 71,
Proteins, of membranes, 5, 6, 10, 36, 43
Protoplasts, *see also under name of bacterium concerned*,

Protoplasts—*continued*
  antibacterials, and, 72—76
  methylglucoside uptake by ghost membranes, 90
Pseudomonas aeruginosa, antibacterials and, 72, 76, 77
P. natriegens, 102

R

Refsum's disease, 37

S

Saccharomyces cerevisiae, see also under Yeast,
  plasmalemma of, 41, 42, 46, 47
  transport carriers of, 100, 103, 104
Saligenin, 132, 133
Salmonella typhimurium,
  sugar transport in, 95, 105
  sulphate transport protein of, 101, 103, 104
Sensor groups, 82, 84
Small intestine, *see also* Intestine,
  sugar transport in, 109—116, 117—130
Sodium, *see also* Na$^+$,
  azide, 93, 94, 95
  fluoride, 94, 95
Sorbitol transport, 95
Spheroplasts, 39, 42, 76, 77
Sphingolipids, 36
Staphylococcus aureus,
  membrane-active antibacterials, and, 71, 72, 77
  sugar transport in, 95, 105
  wall peptidoglycan and membrane of, 61, 64, 65, 66, 68
Sterols, 42, 44
Streptococcus, peptidoglycan of, 61
Streptomyces,
  D-alanine carboxypeptidase of, 65—69
  peptidoglycan of, 61, 65
Submitochondrial particles, 18—32
Substrate-binding sites, *see also under names of specific compounds*, 101—102, 109, 110
Succinic dehydrogenase, of membranes, 52—57

Sucrase,
  antibodies of, 125
  intestinal sugar transport, and, 117—130
  sodium activation, 120
Sucrase-isomaltase complex,
  antibodies to, 127
  glucose carrier as, 128
  glucose fixation by, 120—128
  isolation of, 120—121
  sugar binding site in, 117—130
  urea plus trypsin effect on, 123
Sugar(s), *see also under specific names*,
  glucose fixation, effect on, 123, 124
  transport,
    amino acids, and, 132
    antibodies, and, 125
    bacterial, 99—103
    carrier proteins for, 99—103
    6-deoxy-L-galactose, and, 110—114
    diagram of steps in, 115
    intestinal, 109—116, 117—130, 131—139
    phlorizin, and, 131—139
    phosphorylative, 87—97
    sodium activated, 110—115, 120
    sucrase, and, 117—130
Sulphate, transport protein, 101, 103, 104

T

TCC, *see* Trichlorocarbanilide
TCS, *see* Tetrachlorosalicylanilide
TDG, *see* Thiodigalactoside
"Tectins", 6
Tetrachlorosalicylanilide, (TCS), 71—77
Tetrapeptide subunits, of peptidoglycans, 59—63, 64, 65
Thiodigalactoside, (TDG), and transport, 96, 101
Thiogalactoside transacetylase, 99
Thiomethyl β-galactoside, (TMG), 96
Thyroxine, and mitochondrial membranes, 28—30, 31, 32
TMG, *see* Thiomethyl β-galactoside
Tocopherol,
Translocation, *see* Transport
Transmural potential difference, 112, 113

Transpeptidases, in peptidoglycan biosynthesis, 65, 68, 69
Transport, see also under name of compound transported,
  amino acids, of, 131–139
  mechanisms of, 81–139
    conformational rearrangements and, 110, 114–115
    models of, 105–106
  sugars, of, see also under Sugars, 87–97, 109–116, 117–130, 131–139
Transport proteins in single cells, 82, 87
  L-arabinose, for, 102
  galactose, for, 100, 101, 103
  β-galactosides, for, 99–100, 103
  glucose, for, 101, 103
  glucose-6-phosphate, for, 102
  identity of, 105–106
  isolation of, 99–106
  L-leucine, for, 101, 103
  mannitol, for, 102
  properties of, 103
  purified, 103, 105
  substrate binding by, 101–102
  sulphate, for, 101, 103
  tests for, 102–104
Trichlorocarbanilide, (TCC), membrane-active effects of, 71–76
Trilamellar membrane structure, 5, 10, 11, 12
Tris, and glucose binding, 121, 122
Triton X-100, and sugar uptake, 120, 121
Trypsin, effect on membranes, 10–12, 16
Tubules, of mitochondrial membrane, 19, 20, 21, 25

U

Uncouplers, 93
"Unit membrane", 4
Unsaturated fatty acids, 40
Urea-trypsin, and glucose binding, 123
Urographin, 40, 44, 45

V

VAL, see Valinomycin
Valinomycin, membrane activity of, 71–76
Vancomycin, 68

W

Wall, in relation to membrane, 39, 40, 47, 48, 59–70, 77
Welchia perfringens, peptidoglycan, 61

X

X-ray diffraction studies, of membranes, 5, 9–12
D-Xylose, and glucose-fixation, 123, 124

Y

Yeast, see also *Saccharomyces cerevisiae*,
  plasmalemma, 39–49
  cell wall, and, 39, 40, 47, 48
  chemical composition of, 42–44
  enzymes of, 44–46
  fine structure of, 40–42
  glycolipids of, 47
  isolation of, 39–40
  subfractionation of, 46–47

RETURNED

THE LIBRARY